Wildern
Tapestry

WILDERNESS

TAPESTRY

An Eclectic Approach

to Preservation

Edited by Samuel I. Zeveloff,

L. Mikel Vause,

& William H. McVaugh

University of Nevada Press

Reno • Las Vegas • London

The paper used in this book meets the requirements of American National
Standard for Information Sciences—Permanence of Paper for Printed Library
Materials, ANSI z39.48-1984. Binding materials were selected for permanence
and durability.

Library of Congress Cataloging-in-Publication Data

Wilderness tapestry : an eclectic approach to preservation / edited by
Samuel I. Zeveloff, L. Mikel Vause & William H. McVaugh.
p. cm.
Includes bibliographical references and index.
ISBN 0–87417–200–4 (alk. paper)
1. Nature conservation—Philosophy—Congresses. 2. Wilderness areas—
Congresses. 3. Ecology—Philosophy—Congresses. I. Zeveloff, Samuel I.,
1950– . II. Vause, L. Mikel (Lawrence Mikel), 1952– . III. McVaugh,
William H., 1939– .
QH75.A1w5491992
333.78'2—dc20   92–19508
CIP

University of Nevada Press, Reno, Nevada 89557 USA
Copyright © 1992 by University of Nevada Press
All rights reserved
Book design by Kristina E. Kachele
Printed in the United States of America
2 4 6 8 9 7 5 3 1

Dedications

To Linda

S I Z

To my wife Janis; my daughters Kelly, Emily, and Sarah;
my son Jared; and my mother Stella B. Vause

L M V

To my son Nate, and my daughters Lisa, Becky, and Tasha

W H M

Contents

Preface

NORTH America's Intermountain West must spur many to think about wilderness. A land of stunning beauty, it harbors much that nature lovers hold dear: majestic mountains, rugged canyons, stark deserts, and an unusually diverse flora and fauna. The region is rich with both designated and proposed wilderness areas, and is currently the focus of intense political debate. It was thus natural for this locale to become the site of an interdisciplinary conference on wilderness.

In 1988, the three editors of this volume, all professors at Weber State University in Ogden, Utah, conceived of the North American Interdisciplinary Wilderness Conference. Our hope was that such an approach would encourage individuals to make connections with those in other disciplines. A primary goal was to further our understanding of the many meanings and values of wilderness, which could consequently provide an increased justification for preservation of these areas. The first three annual conferences, in February of 1989, 1990, and 1991 have each been rewarding. Keynote speakers have included Chris Bonington, David Brower, Cheryl Charles, Maurice Hornocker, William Kittredge, L. David Mech, Michael Soulé, and Richard White; several of them have contributed to this book.

This volume, which largely emanated from the first conference, has undergone a gradual, if at times, a punctuated evolution (apologies to Eldredge and Gould 1972). Our aim was not simply to compile papers from disparate disciplines, but to present a coherent collection on stimulating wilderness issues. The book was further shaped by inviting complementary contributions. Ultimately, the conference's finest submissions plus these solicited papers formed this volume.

Many people assisted with its production. We first thank those who had the intellectual curiosity to participate in the first conference in 1989. All subsequent meetings and related publications are indebted to them. Many at Weber State have supported this conference: Stephen Nadauld and Paul Thompson, its past and present presidents; Robert B. Smith, Vice President for Academic Affairs; the Centennial Scholarly Activities Committee; Associated Students of Weber State University; and several deans and department chairs: Sherwin Howard, Richard Sadler, Cy McKell, Candadai Seshachari, and Richard Grow. Bonnie Donaldson, secretary for the Department of Zoology, cheerfully processed too many manuscript drafts. Samuel I. Zeveloff worked on the book's final editing while on leave with the Large Animal Research Group of the Department of Zoology at the University of Cambridge, England. Finally, Thomas R. Radko, Director of the University of Nevada Press, was always a strong source of encouragement as he moved the volume along during a thorough review process.

Various individuals provided insightful chapter reviews. Those permitting us to reveal their identities are Skip Ambrose, Lee Aulman, Mark Boyce, William Burbridge, Dix Cloward, Rosemary Conover, Pierre Crosson, Paul Groth, Kristen Hawkes, Judith Heintz, David Iverson, Gary Miller, Con Schallau, Glenn Steward, and Steven Strom.

A version of Jay Hair's paper was originally published in the University of Idaho Wilderness Distinguished Lectureship Series. We are grateful to James Fazio of that university's Department of Wildland Recreation Management for enabling us to include it. The chapter by L. David Mech first appeared in a volume entitled *Working Papers, The President's Commission on Americans Outdoors*. "Burrowing Owls" by Terry Tempest Williams appears in revised form in her collection *Refuge: An Unnatural History of Family and Place* (Viking, New York); "Yellowstone in Winter" by William Kittredge is from *Owning It All* (Graywolf, St. Paul, Minnesota); "The Arctic Desert" by Howard McCord is from *Walking Edges* (Raincrow Press, Ohio); and Ann Ronald's and Russell Burrows's papers were originally published in *Western American Literature*. We are thankful to them for granting us permission to include their works.

Finally, we are grateful for the lasting contributions of those from the diverse disciplines that constitute wilderness studies. Our thoughts, interactions, and efforts to preserve wild areas have benefitted from these individuals. We hope that this volume similarly stimulates others to appreciate and understand the rich meanings of wilderness and to promote its preservation.

Samuel I. Zeveloff, L. Mikel Vause, William H. McVaugh
Ogden, Utah
January 1992

Reference

Eldredge, N.; Gould, S. J. Punctuated equilibria: An alternative to phyletic gradualism. In: Schopf, T. J. M., ed. Models in paleobiology. San Francisco: Freeman, Cooper, and Company; 1972: 82–115.

Introduction

&

Philosophical

Underpinnings

A Multidimensional Mandate
for Wilderness Preservation

SAMUEL I. ZEVELOFF

L. MIKEL VAUSE

WILLIAM H. McVAUGH

PROTECTION of natural areas in the United States began at least as far back as 1872 with the creation of Yellowstone National Park. Setting aside wilderness—areas essentially free from development and thus the amenities of national parks—did not begin until some fifty years later. In 1924 the pioneering ecologist Aldo Leopold orchestrated the establishment of our first official wilderness: 574,000 acres of New Mexico's Gila National Forest. It seems that his primary motivation was to ensure the survival of pack-trip hunting experiences (see Flader 1974). Others instrumental in developing a rationale for wilderness preservation include David Brower, Arthur Carhart, Robert Marshall, and Howard Zahniser (e.g., Baldwin 1967).

The enactment of the Wilderness Act in 1964, some twenty-eight years ago, rewarded their efforts. Through this act, a National Wilderness Preservation System protecting over 9 million acres of national forest land was established. Formally designated wilderness in the United States now totals some 474 separate areas and over 91 million acres—an area roughly the size of Montana. Over half of the acreage lies in Alaska, largely the result of the Alaska National Interest Lands and Conservation Act of 1980.

We have entered what is likely to be the final chapter of determining which lands will remain as wilderness. Currently, about 25 million acres involving some 860 new areas are under consideration for wilderness designation (Leibman 1989). Virtually all are administered by the Bureau of Land Management (BLM); lands under this agency's jurisdiction have not previously been so considered. Most of the BLM's 175 million acres in the lower forty-eight states are arid, desert country. But in addition to deserts, proposed BLM wilderness bills seek to protect canyon country, mountain ranges, and even an old-growth rainforest in Alaska's Tongass National Forest.

Arguments for wilderness preservation usually center on the importance of maintaining remnants of Earth in a natural, pristine state (e.g., Bainbridge 1984). This becomes an ever more urgent task, as habitat destruction proceeds at a frighteningly accelerated pace. Thus, a compelling rationale for leaving natural habitats undisturbed is to halt the rampant human-induced extinction of animal and plant species. In fact, it may only be possible to protect many species, particularly large wide-ranging ones, within the confines of sizable natural areas. The vital necessity of preserving biodiversity, the richness of life on this planet, has been effectively articulated for a variety of ethical as well as economic reasons (e.g., Ledec and Goodland 1988; Wilson 1988; Ehrlich and Wilson 1991).

Saving the world's unique physical environments and life forms will undoubtedly remain as the underlying premise for wilderness preservation efforts. But current thinking promotes wilderness as simply an inextricable part of being human. Of course, people first existed in a "wilderness," and the desire to be within a wilderness might even have a genetic basis. At any rate, features of wild areas seem indelibly etched in our psyches. Wilderness themes commonly appear in the fine arts, literary analysis, prose, and cultural studies. The case for wilderness preservation might then be based on the ways in which it is a profound, if not an essential, component of our existence.

An examination of some important philosophical and conceptual aspects of wilderness follows this introduction and provides a useful groundwork for the chapters that follow. They justify not only the desirability of preserving existing wilderness, but also argue for restoring additional areas to their wild origins.

Many of the ways in which we view wilderness can be traced to the works of writers, poets, and painters. Ralph Waldo Emerson, Henry

David Thoreau, and John Muir are but a few of those who have shaped our wilderness consciousness. The volume's next section, "A Heritage of Wilderness Literature," illustrates the yet pervasive influence of this discipline.

Fine nature writing continues to flourish. We are delighted to offer work by some of the best writers anywhere. The authors in the section, "Original Wilderness Prose," have proven to be particularly adept at using natural history themes. We hasten to point out their other works cover a lot of territory, not all of which is centered on wilderness. All should become familiar with their art.

Thinking about wilderness topics leads to interesting speculations. In the section "Historical and Contemporary Societal Views," we observe that wilderness is indeed a slippery concept. Provocative examinations seem to reinvent the very meaning of wilderness. Authors in this section pose such questions as "Are there connections between wilderness and feminism?"; "Can urban areas harbor wilderness?"; and ultimately, "Is wilderness simply an invention of the human mind?".

Establishing and maintaining wilderness involves substantial management. Areas most suitable for wilderness designation must be determined, wilderness systems designed, and negative impacts modified. Many papers (e.g., Hendee and Schoenfeld 1978) and even entire conferences (e.g., Lucas 1986) have been devoted to such topics. The section "New Management Concepts" offers innovative ideas concerning education and economics.

"Future Paths" provides a glimpse of where we may be headed. Jay D. Hair, president of the nation's largest conservation organization, the National Wildlife Federation, provides a compelling discussion of wilderness concerns. Finally, wolf biologist L. David Mech, the individual perhaps most closely identified with wilderness ecology research, offers his views on our relationship with nature.

This book explores a wilderness with many meanings and values. It takes us from our roots as sentient creatures to where we are going and what we might become. We offer that we might not reach our potential or maintain our identity without wilderness, a view supported by the essays in this volume. We thus present this book not simply as an interdisciplinary approach to, but as a multidimensional mandate for, wilderness preservation.

References

Bainbridge, B. Management objectives and goals for wilderness areas: Wilderness areas as a conservation category. In: Martin, V.; Inglis, M., eds. Wilderness, the way ahead. Forres, Scotland: Findhorn Press; 1984:114–124.

Baldwin, D. N. Wilderness: Concept and challenge. Colorado Magazine 44:224–240; 1967.

Ehrlich, P. R.; Wilson, E. O. Biodiversity studies: Science and policy. Science 253:758–762; 1991.

Flader, S. L. Thinking like a mountain: Aldo Leopold and the evolution of an ecological attitude toward deer, wolves, and forests. Lincoln: University of Nebraska Press; 1974.

Hendee, J. C.; Schoenfeld, C. A. Wilderness management for wildlife: Philosophy, objectives, guidelines. Transactions of the 43rd North American Wildlife and Natural Resources Conference 43:331–343; 1978.

Ledec, G.; Goodland, R. Wildlands, their protection and management in economic development. Washington, DC: The World Bank; 1988.

Leibman, D. In celebration of wilderness! Conservation 89 7(9):7–10; 1989.

Lucas, R. C., compiler. Proceedings—National wilderness research conference: Current research. Ogden, UT: U.S. Dept. of Agriculture Forest Service General Technical Report INT-212; 1986.

Wilson, E. O., ed. BioDiversity. Washington, DC: National Academy Press; 1988.

The Barbarian Link

BROOKE WILLIAMS

Y E A R S ago I was flying from New York to Salt Lake at night. It was clear and even from 33,000 feet I could see the lights of the towns and cities below. The further west we got the more dark space existed between the small outposts. In a sense, the lights meant protection; a comfortable house, with lamps, a stove, soft bed, and convenience. Beyond the light, who knows? To modern man, the lights seem to form a border separating what is acceptable from what is not. Urban people, accustomed to the womb that has been built for them, go beyond this border only when the benefits seem to outweigh the comforts that will need to be sacrificed.

Ages ago, those who lived on that border, at the edge of society, and took from it what they could were called barbarians. Originally, barbarians were looked down on, considered less than human by the Greek societies they preyed on. Actually, barbarians were ostracized and forced to become outsiders simply because they could not speak Greek. Since then many cultures have had their barbarians to contend with and now the word brings up images of huge, warlike men with bulging muscles who would as soon pull off another man's arm as give him the time of day.

The barbarian concept has undergone considerable study. Walter J.

Ong in his book, *The Barbarian Within* (1962) has a chapter entitled, "Wilderness and the Barbarian." In it he says,

> For the advance of civilization barbarians may be necessary. . . . no culture is worth preserving exactly as it is. All cultures need improvement, and need it rather badly and the operations of barbarians on a culture, whatever the immediate effects, can result in a sorting out of what is valuable from what is not.

He is talking about how, in the long run, these outsiders may have done more good than harm. In the course of history barbarians may have strengthened society by infusing it with diversity, making it aware of its weaknesses, and keeping it alert. It was not predation by barbarians which society feared as much as the inflow of radical and threatening ideas, the infiltration of individuals planting seeds of doubt, encouraging spontaneity and giving value to rebellion.

A modern vision of barbarian would be difficult except that I have managed a new definition for the word. I feel it is fair to do this as the dictionaries make no mention about modern barbarians. My definition will be true to history in that the new barbarians will be in some sense, outsiders. And I doubt that they speak Greek, Latin maybe, but only in naming plants and animals.

New barbarians too are outsiders in the sense that physically they would rather be under stars than roofs. Emotionally, they are on the outside of modern society because they react to rock and sky rather than money and comfort. They have a two-way relationship with the earth and are not content to sit on a flat rock and let nature speak to them. It is an active relationship, a dance wherein the combination of person and planet, pulsing muscles, stretching joints, fully inflated lungs, and skin moistened slightly by sweat are catalysts often causing extraordinary events.

Whether they are hiking endless arroyos, skiing bottomless, back-country powder, climbing granite cliffs, being blown around a lake's surface on a sailboard, or dancing with a campfire's flames, new barbarians are discovering an unnamed part of themselves which is difficult to describe.

There are writers who have referred to it. Terry and Renny Russell (1969) wrote in the first Sierra Club book I ever owned, *On the Loose*:

"Even if our species has lost its animal strength, its individual members can have the fun of finding it again."

Black Elk, a Sioux holy man whose life straddled the turn of the century was quoted in John Neihardt's (1961) book, *Black Elk Speaks*: "It may be that some little root of the sacred tree still lives. Nourish it then, that it may leaf and bloom and fill with singing birds."

And the poet, Mary Oliver (1986), says in *Wild Geese*:

> You do not have to be good.
> You do not have to walk on your knees
> for a hundred miles through the desert, repenting.
> You only have to let the soft animal of your body
> love what it loves.

Animal strength, the root of the sacred tree, the soft animal of your body, loving what it loves—all these describe the part of us that we share with all humans, not just here and now, but forever and wherever.

Ten thousand years ago, the sacred tree Black Elk spoke of flourished in every person who walked the earth. Their lives depended on it. It was the instinct that let them know how to hunt and where water could be found. It showed them direction and gave them insight. It gave them knowledge about plants and animals. It told them when to pick up their small bundles of belongings and move on.

Their world was physically no larger than the space they wandered within. Emotionally, it expanded to encompass whatever higher reason required as explanation. Their space was the source of the myths that surrounded them.

These people, the hunters and gatherers, the archaic nomads, were physically no different than we are. Evolution in full strength does not change a species as sophisticated as ours in such a short time. How then, can this wisdom, that seemed to come through the combination of a natural course of evolution and unencumbered space, be neglected in our quest for survival? One reason why evolution works is that only the fittest members of a species survive. It is hard for many moderns to admit, but humans are humans because of the concept of survival of the fittest, in conjunction with the ability of a species to change. Modern religion has led us to believe that we are the ultimate species, that we are above the laws of evolution. When the truth is known, we may discover that our sacred technology has thrown a wrench into a more incredible plan for perfection than we have the capacity to dream. We might find that our physical potential has not been met. Because of this cosmic wrench, survival of the fittest has become survival of the masses, giving everyone the inherent right to live regardless of what skills or knowledge they have developed, regardless of what deformities or defects they may have. We are no longer required by the demands of nature to find our own food or heal ourselves. We see no need to limit our population, predict the weather, or find our own way. Society and technology do it all for us.

Have we lost the ability to save ourselves, or is this power the root of the tree that lives in various stages in each of us? Carl Jung, with his concept of archetype, maintained that we share our dreams and symbols with every man and woman who ever walked the earth. Could it be that we also share with them a spark of life that holds keys to our survival, something that proves beyond any doubt that we are integral and subject to all the forces, named and unnamed, tamed and untamed, that control all life—higher and lower?

Sally Cole is a barbarian. She is an archeologist. She spends months every year wandering in redrock canyons looking for remnants of ancient cultures. She will walk for days to a site where an Anasazi shaman painted figures on the rock face of a protected alcove. She will look at the wall for days, measuring, mapping and diagraming it in her notebook. Each day she sees more until eventually she may find the path back through the creative process to the source that inspired the art. She knows that she and the artist have much in common. She has no need for something to believe in, but is continually amazed by events that support what she instinctively knows. If you blindfold her and turn her around and around and around then let her go, she will head for her secret, wild place. At home, she will watch the evening news and care only about the weather.

About ten thousand years ago people began to see the value of planting seeds, rather than moving about seasonally to find them. Much has been written about the onset of agriculture and the demise of the nomadic hunter-gatherer. It is difficult to see this new path as wrong, especially considering how agriculture gave way to civilization and comfort. But, the astounding, eye-opening fact is that as a species, we no longer have the innate ability to adapt to our changing surroundings. We are now able to adapt the surroundings to meet our needs. Our future now depends on the limited understanding and knowledge, not to mention the ego, of those we choose to control our lives.

The ability to grow and store food diluted the relationship between man and nature. Imagine the early generations who had discovered certain hybrids of seeds which could be planted to produce food crops. Was there any uneasiness that first season when they stayed to wait for their crops to grow? At first, was it strange to have accumulated more than they could carry? The details are sketchy, even to the imagination.

Does the man with a warm house and a bin full of extra food become one of his own gods and a source for his own mystery?

This cycle toward increased comfort is a marked diversion from natural evolutionary tendencies. Why the change? Was the move towards agriculture an evolutionary one? If this were the case, one would think that societies could live in closer harmony with one another and with nature. The discrepancy has to do with the difference between nature and culture. Agriculture was the dawn of culture and marked the rapid independence from nature. Cultural evolution has changed the world a hundred times faster than biological evolution. Our bodies as we know them were millions of years in the making. Our culture, ten thousand. The danger is that the two forms of evolution don't work at the same rate. Our bodies can't adapt fast enough to the different environments that society and technology are constantly creating.

What is our true nature? Is it based on modern culture even though it has been in the making such a short time? Or is it in our deep core formed from the millions of years of evolution? A similar problem is familiar to any scientist studying human behavior. Are our actions the result of the environment we live in or the genetic information contained in our cells? The most common consensus is that the genes must be there to begin with or the environment has nothing to work on. Genes can only be encouraged or discouraged. What about the genetic reality of our true nature? Are there genes for humanness hanging like Christmas tree lights from one of our forty-six chromosomes that no one has bothered to name? Or is it something much deeper—something that our bones are made up of, that gives color to our blood, and sparks the pumping of our heart? Thoreau said that a man must live according to his true nature or he will die. Our true nature lies deep inside us, covered with layers of thought and habit, remnants from the fast and furious growth of society, entangled in a web of influence that says that success is defined by the individual and is no longer measured by the species as a whole.

Creighton King is a barbarian. Running is part of his true nature. He knows because he has spent fifteen years on the trails of western mountains, trying to discover the limits of his own body and imagination. He has found that in running his body reacts rather than responds, that he is capable of spectacular things which are not limited by thought. He is confident because he has passed the simple test of saving himself. He knows things that have no words, things that transcend teachers and books and Sunday school—the kind of knowledge that stands up on its hind legs and doesn't need propping up by understanding. He just knows. It's personal.

If the millions of years that man and woman have been walking upright on the earth could be seen in perspective, our modern society might be no more than an opening in the eye of time. It has opened quickly. Would it be absurd to think it could close even faster? That our entire existence would be a blink in time? Inner city decay, global overpopulation, the itch for absolute knowledge and power by many different factions, and a technology with the power to destroy all life suggest this possibility. Those we have placed in control of our path have two choices. They can continue in a specialized and reductionist direction where most available money and talent goes toward the creation of unnecessary comforts for those who can afford them and for fueling an arms race with weapons we hope never to use. Or they can look far enough ahead and see that reductionism can go only so far and begin looking for the interconnectedness of everything.

What can we hope for? A forest of sacred trees in the form of people in whom the root has been allowed to grow by a society that sees the need to preserve and encourage the blossoming of innate sensibilities.

I am talking about wildness.

Wildness is a state of mind. Remember that Thoreau did not say, "In wilderness is the preservation of the world." He said "In wildness." But wilderness must be protected as a place where wildness is allowed to happen, where wild, evolutionary tendencies are encouraged. In this day and age when pressures on the land are severe and time is money and space means planets and stars, not open country, wild space must be given priority and value in terms different than dollars. We are slowly moving away from Freud's idea that since science can be measured and proven that it and it alone should provide the basis for world view and philosophy of life and that man's true nature can only be properly explained through scientific inquiry (see Hall 1978).

Jeff Foott is a barbarian. He is a wildlife photographer. His intense scientific inquiry gives him an edge professionally. But his passion comes from the peace and persistence that allows him to share gazes with whales, to live in a tree with an owl, or to share his home with a pocket gopher. He has learned, like his hunting ancestors did, to crawl into an animal's skin. He knows what society has to offer but wants more. He is an outsider because he has replaced the word "success" with "succession;"—the continuation of life with no particular goal in mind, building thoughts, ideas, actions, one on top of another, with each layer getting closer to new truths which reinforce age-old myths and add up to a better understanding of life's actual intentions.

Although Freud's thinking is widely accepted by our modern world, it only goes as deep as our relationship to society, which is only a small piece of the distance we have come. In our culture, science is the only language we speak, the lone instrument we hear. Think about going to a concert and hearing only the cymbals. While they perform a function vital to the success of the symphony, they can't work on their own. The other instruments—myth, magic, movement, art, expression, dance, intuition, personal experience—are all there, deep inside us and have been forever. Freud felt that these characteristics had no use in modern society. What he never considered was that these primitive traits, not science, secured the survival of our ancestors and, who's to say, maybe our own. At least they are part of the path connecting us to all that is human. Freud felt that modern man does not need this connection and is better off without it. Such a connection may feed the Id, his name for the part of every individual's personality interested only in the satisfaction of innate needs. He felt the Id must be controlled for society to function. According to Freud, the Id is the impulsive, instinctive part of a person which is not governed by logic or reason and has no morals or ethics. His ideas about the Id came primarily from psychoanalysis and studying the dreams of patients with neurotic symptoms (see Hall 1978).

Freud's study of mankind left him pessimistic. He felt that the internal forces of the Id were so strong that the rational forces added to the personality by the laws that govern civilization were no match. The result is a society formed largely on man's irrationality. Could it be that Freud's negative opinion of the Id came from looking into the lives of restrained patients who were out of control and dangerous? What would his opinion have been if he had studied healthy people who allowed their Id to wonder and wander without restraint? What if instead of the rapist, he had looked into the life of the mountaineer or the artist? Instead of the murderer, the poet? (See Hall 1978.)

Is it possible that Freud's Id and Mary Oliver's (1986) soft animal are the same thing? Is this the sacred tree that is nurtured by modern barbarians? I think so. I am hopeful.

So, barbarians, who are you?

You are doctors, lawyers, waiters, shopkeepers. You are chemists and teachers, artists and athletes. You are plumbers and carpenters. You work for Thiokol, Hercules, even Kennecott, and U.S. Steel. You work for the federal, state, and local governments. You are good at what you do because what you know gives you higher reason. Whoever you are, you have tapped into a piece of what life requires.

Go to a public land hearing. Notice how those who want to develop and those who want to save seem to speak a different language. The day may come when those who care about wildness will be on both sides of the issue. And what will be at stake will not be the land, but the method in which to hold it in the highest regard. Both sides will be speaking the same language. It will not be Greek.

So think about it. Think about the barbarian part of you, that small animal inside you. Think about the wild land and how you are better because of it. How those around you at work and home or passing you on the street are better because of you.

So even if your ancestors didn't hunt and gather on this continent, you can; maybe not seeds and berries, but insight, feeling, and vistas to carry back in your mind. Strive to be native. The world will benefit.

If wilderness needs to be measured in human terms, and unfortunately, its survival requires that it is, then let it be seen as the promised land. Let it be known as the space where the secrets are kept and the clues to our own well-being are hidden. Let it be known as the seed source for the hybridization of science and creativity, for the germination of higher reason and the growth of hope. At least, let it be seen as the place where the barbarians go to discover life the way primitives did, to learn requirements for our species' survival, in case modern society and civilization forget.

References

Hall, C. S. A primer for Freudian psychology. New York: Octagon Press; 1978.

Neihardt, J. G. Black Elk speaks. Lincoln: University of Nebraska Press; 1961.

Oliver, M. Dream work. Boston: Atlantic Monthly Press; 1986.

Ong, W. J. The barbarian within and other fugitive essays and studies. New York: MacMillan; 1962.

Russell, T.; Russell, R. On the loose. San Francisco: Sierra Club Books; 1969.

Wilderness Restoration
A Preliminary Philosophical Analysis

WAYNE OUDERKIRK

Introduction

T H E six-million-acre Adirondack Park in northern New York State incorporates a unique mixture of private and public lands. Many sections are multiple-use areas, but large portions are designated forever wild by the state constitution. In the nineteenth century, most of the region was heavily logged; mining was also common. Since the forever-wild area was created in 1885, much of its wilderness character has gradually returned.

But not completely. Moose, lynx, wolves, peregrine falcons, bald eagles, and cougars all became extinct in the region. The moose has begun to return on its own (McKibben 1988); bald eagles, lynx, and falcons have been reintroduced by the State Department of Environmental Conservation. There is serious talk about reintroducing wolves and cougars, in an effort to restore the Adirondack wilderness to something approaching its former character.

The Adirondacks form only one example of an attempt at restoring a wilderness area. Yet despite the promise and hope offered by such efforts, we should examine questions of ethics and values before endorsing restoration unequivocally.

Throughout, this essay considers active restoration of wild places, as opposed to simply allowing nature to restore itself. Using this distinction, much of what today exists in the Adirondacks is the result of letting nature take its course after humans had intervened. Reintroduction of species, however, involves additional human intervention.[1]

Some (Godfrey-Smith 1979; Nash 1982; Rolston 1986) have argued forcefully for preserving existing wilderness in as pristine a condition as possible. They assert that wilderness has value. Hence, they argue, any human activities that destroy such areas should be *prima facie* prohibited. Rather than discussing preservation, however, I want to ask whether we ought to restore damaged wilderness. This exploration looks at two questions. Are there any normative qualifications or limitations on wilderness restoration? And, are there any moral or other imperatives directing us to restore wilderness?

First, although present efforts to restore wilderness areas have value, such efforts ought to consider interrelated issues. These include ecological knowledge, economics, and justice.

Secondly, when swift action is required, a discussion of philosophical niceties may need to be suspended. When an oil spill befouls a wilderness beach and marsh area, for example, there is no question about the right course of action. Here a moral imperative directs us to restore an ecosystem, ensuring it the least possible interruption or loss. Still, even in such a pressing situation, the moral issues of responsibility and justice exist.

Nash (1982) has pointed out the difficulties in defining wilderness. He cites such issues as the subjective and emotive "freight" the term carries, the question of how much human intrusion causes us to deny that a place is wild, and how much area is required. Using Nash's idea (1982) of a spectrum of conditions, ranging from the purely wild to the civilized, wilderness is simply that which falls closest to the purely wild. Thus, some areas considered for wilderness restoration might, at the beginning of the process, be closer to the civilized end of the spectrum. The goal: to move them toward the wild end.

But a related issue arises immediately: would such a restored area qualify as "real" wilderness? This is a pseudo-issue though. For one thing, since humans have affected virtually all of the globe in some way, it could be asked of any presently existing wilderness area. For another, the concept of wilderness is subjective and emotional. So some would reject, and others accept, restored areas as wilderness. Perhaps it simply depends on the closeness of the restored area to the purely wild end of Nash's spectrum.

What about defining restoration? "Restoration aims to reestablish viable native communities of plants and animals" (Wolf 1988, 110). Although not mentioning the goal of remaking wilderness, this definition has the virtue of emphasizing native communities, rather than economically useful end products.

General restoration and wilderness restoration are closely related. But are they identical? Again consider Nash's spectrum of conditions. Many restoration efforts intend "to reclaim biologically" (Cairns 1980). Other efforts have recreational or aesthetic purposes (Berger 1985). If the intent is to approach the wilderness end of the spectrum, then we are dealing with wilderness restoration. Henceforth, unless clarity dictates otherwise, the term "restoration" designates wilderness restoration.

Restoration has roots in venerable conservation practices of land and stream reclamation, reforestation, and even pollution clean-up. Whereas many conservation practices have economic or recreational purposes, restoration implies that the area will become and remain wild.

In Favor of Restoration

Restoration has great appeal. Combining a concern for the original condition of our planet with our technology and rapidly growing knowledge of ecosystems, restoration provides an attractive option for environmental activism. In addition, rapidly disappearing wilderness and individual species provide even greater cogency to the other arguments favoring restoration.

Wilderness restoration, like the restoration of historical buildings, preserves pieces of history (Nash 1982). As with historic restoration, a successful project provides us significant contrast with contemporary civilization. However, if this was the only reason to restore wilderness, it could be argued that existing wilderness preserves adequately fill these roles. But this sort of argument omits or ignores the dynamic nature of wilderness ecosystems. Historical objects are static, though interpretations of their meanings may be dynamic. It would be ill-considered, I think, to base restoration efforts on this sort of justification without extensive, strong qualification of the suggested analogy between historic objects and natural areas.[2]

Restoration also connects our present actions with the rights of future generations. Much has been written on our obligation in this matter (e.g., Feinberg 1974; Partridge 1981). But if present trends in land use and pollution continue, little, if any, wilderness will remain for future generations to know. Assuming that they have rights and at least some interest in common with us (Feinberg 1974), some of them undoubtedly will have an interest in wilderness. While this argument seems to militate more for preservation than for restoration, presumably, future generations would also prefer wilderness to degraded, polluted, or eroded areas. To that extent, the future-generations argument supports wilderness restoration. However, since there is some controversy attached to the notion of rights of future generations (DeGeorge 1979)[3], this argument is not decisive.

In an effort to strengthen and clarify the claim that future generations have rights, Callahan (1971) and others have proposed an intergenerational "contract" in which we, by acting in the interests of our descendants, fulfill a debt or obligation to our own forbearers who, in their own time, acted somehow in our interests (though in the process of doing so, they polluted or destroyed).

By using the concepts of debts and obligations, the intergenerational argument can be interpreted as follows: Since we are all the beneficiaries of activities that have degraded the earth (i.e., we have incurred a debt), and since we have it in our power to correct some of the damage (i.e., to repay the debt), we have an obligation to do so. So while the vast majority of the benefits we have received from our forbearers are not specifically in the form of pristine wilderness, we have the opportunity to confer that benefit on our descendants. This argument also can be used to support either restoration or preservation.

An idea favoring, but not establishing, an obligation to restore is that of know-how. Our increased knowledge of biology, especially of ecosystems, allows us to preserve nature more wisely. Although this knowledge was not necessary for effective preservation, it permits restoration of degraded ecosystems. Knowledge of an organism's niche, for example, enables us to make better decisions about reintroducing it and about its chances of becoming reestablished.

The technology available for restoration has changed so drastically that we can accomplish formerly imposing or impossible tasks. As one simple example, it once took five months of manual labor to reseed and replant a marsh area several hundred feet wide, a job currently requiring only five days with appropriate equipment (Berger 1985).

Increased knowledge and ability do not, in themselves, imply any obligation to restore. However, a traditional condition on obligation advanced by ethical theory is that "ought" implies "can." That is, if we are able, we ought to restore. On the other hand, an alleged obligation to do something is cancelled, or at least postponed, if we are unable to do it. So if other arguments favoring restoration establish a *prima facie* obligation, the ought-implies-can condition obliges restoration.

The amount of wilderness left on the planet and its prospects for survival indicate that we ought to restore. Wilderness presently comprises only two percent of the contiguous forty-eight United States (Rolston 1986). In the early 1980s tropical forests disappeared at the rate of 11.3 million hectares per year (Postel and Heise 1988). And the pressures for more development continue to grow. Witness the pressure from oil companies to drill in the Arctic National Wildlife Refuge.

In the face of these and other arguments, many have offered reasons to preserve wilderness (e.g., Godfrey-Smith 1979; Nash 1982). These reasons include the scientific, agricultural, recreational, religious, and aesthetic benefits of wilderness. These same reasons can be used to argue for the ethics of restoring wilderness. After all, if such a valuable part of the world is disappearing, and we have it in our power to restore some of it, then doing so honors and reestablishes the benefits that wilderness confers.

Clearly, people profit from aesthetic and scientific benefits. However, if nature itself has value, before being interpreted in terms of human interests and experiences, that would provide a powerful argument for restoration. Rolston (1986) has argued for just such a thesis:

> A tough-minded analyst can insist that nothing has value until humans arrive. One moment that may seem right, but then again is he not a provincial who supposes that his part alone in the drama in which he participates establishes its worth? (70)

Nevertheless, traditional value theory seems to reduce all value to human interests. If we take ecology and evolution seriously, however, and accept that humans are part of the ecosystem, then saying that values are created by human interests makes values something different or separate from the ecosystem. Such a theory contradicts what science tells us (that the earth is a system in which everything is connected) and what value theory tells us (that values literally are found only in human reflection). In Rolston's theory, values are in nature.[4]

Rolston's theory supports the importance of wilderness restoration. He argues that if nature has value in itself, then restoring an area to its natural state is a worthwhile activity because its goal is intrinsically valuable. Thus, if we contribute to the degradation of wilderness, we have an obligation to repair it, to pay restitution, as it were.[5]

Finally, Wolf (1988) based his argument favoring restoration on the critical situation regarding the extinction of species. Previous species extinctions did not likely result from the actions of a single species. Yet now the actions of humans cause many extinctions, and the situation worsens. Wolf (1988) cites predictions that fifteen percent of all plant species, twelve percent of bird species, and a perhaps unknowable number of insect species in tropical rain forests are likely to become extinct by the end of this century if present trends of deforestation continue.

Deforestation creates a patchwork of small pockets of forest, unused by loggers or farmers and insufficient to support the original forest's species diversity. Wolf (1988) advocates connecting the pockets to restore larger areas and prevent at least some extinctions. Moreover, we cannot wait for nature to heal itself to prevent these extinctions: The fragile rain forest ecosystem might take up to 150 years to recover completely from slash-and-burn farming. By that time most of the feared extinctions would have occurred.

But ought we prevent such extinctions? Obviously, Wolf thinks so. He bases his plea for restoration of crucial habitat on the basis of human interest:

Because so many of the species facing extinction are completely unknown, their biological importance remains a mystery and their potential value to society, an open question. . . . The medicines and foods they may have harbored will never be known. . . . Since so little is known about the earth's biological fabric, the consequences of losing biological diversity cannot be forecast with confidence. (Wolf 1988, 104–105)

He argues that we ought to restore crucial habitat for endangered species, since a few of these species may have value for humans.

Rolston (1986), on the other hand, argues that species have value in themselves, independent of human interests in them: "A species is a coherent, ongoing form of life expressed in organisms, encoded in gene flow, and shaped by the environment" (1986, 210). According to Rolston, its function in an environment gives a species value:

> What is valuable about species is not to be isolated in them for what they are in themselves. Rather, the dynamic account evaluates species as process, product, and instrument in the larger drama toward which humans have duties, reflected in duties [to] species. (Rolston 1986, 218)

Thus, we ought to preserve species because of their role in an intrinsically valuable process. Since species and environment are inseparable parts of the same thing, our duty to preserve the one is inseparable from our duty to preserve the other.[6]

Although Rolston does not discuss habitat restoration *per se*, his and Wolf's positions combined with Wolf's facts argue for restoration of damaged habitat: Such restoration would simply be the means whereby we carry out our duty to endangered species. So the two arguments, Wolf's based on human interests and Rolston's based on the intrinsic value of species, both support restoring damaged ecosystems to preserve endangered species.

These arguments to restore wilderness areas are equally compelling as arguments for preservation. After all, to preserve endangered species why not preserve their habitats before they are harmed, rather than to try to reconstruct them from varying degrees of destruction? Of course, Wolf is arguing from the standpoint of already damaged ecosystems— a perspective that works for restoration. But in looking at endangered species in less threatened habitats (e.g., the moose in the Adirondacks) and in trying to devise an overall strategy to maximize species diversity on the planet, the arguments become cogent ones for preservation.

Complicating Factors

Despite the limitations noted, the reasons favoring wilderness restoration—especially those concerning future generations, the intrinsic value of wilderness, and species extinctions—give us *prima facie* reasons to assert a moral obligation to restore wilderness. Even so, factors exist that have the potential to either restrict or postpone that obligation.

The first set of such considerations concerns the species which live, or lived, in an area to be restored. As a result of human interference, many indigenous species have disappeared. Some restorationists wish to bring them back, to return the ecosystem to its state, as much as possible, before disruption. But such reintroduction might not be fair to the animals.

An ongoing philosophical debate concerns animal rights (e.g., Singer 1975; Fox 1978) and the role of animal rights in an environmental ethic (Callicott 1980).[7] But a commonsense fairness to animals applies, whatever the outcome of these debates.

Reintroducing a species if its niche in the ecosystem has effectively disappeared might not be fair. An ecological niche, of course, "includes all [of a species'] interactions with the physical environment and with other organisms that share its habitat" (Turk 1985, 31). After it has been missing for a time, a species' niche might have been taken over by other species, or the whole system might have changed so significantly that the niche no longer exists. In either case, to release animals into such a community would be unfair; they could not survive without a sufficient amount of territory, sufficient prey or forage, and the absence of excessive competition.

Ecologists are well aware of these factors, of course, but many others who favor restoration efforts are not sufficiently sensitive to them and can bring pressure for quick action when more study is needed. Moreover, enthusiasm for a restored area or for a particular species might blind anyone to potential difficulties. Whether based on the rights of animals or on the obligation not to inflict cruelty, the need for ecological knowledge obliges us, morally, to resist reintroduction of species when such knowledge is missing or incomplete.

Related to this issue, indigenous species might have stepped into the animal's niche. Is reintroduction of the original species fair to a newcomer that moved into its niche? The newcomer was not a part of the original ecosystem, but does that entitle us either to destroy its niche or to eliminate it? As an example, landlocked salmon were fished out of Lake Champlain years ago. Meanwhile, lampreys have invaded the lake via man-made canals. Efforts are underway to reestablish the salmon, but the lamprey, which preys on fish, might prevent this. As a result, many want to poison the lampreys to eliminate or reduce pressure on the salmon (New York State Department of Environmental Conservation and U.S. Fish and Wildlife Service 1987). Aside from the issues lampricides themselves raise, is this a fair process? Many would balk at applying the concept of fairness to the lamprey, but such issues are not restricted to indigenous animals we deem pretty or useful. That it's-ugly-so-let's-get-rid-of-it attitude has led to so much destruction of wilderness already.

A second question relates to the need for ecosystem knowledge as well as human interest and activity. Worldwide, in most potential wilderness areas, humans are periodic or constant visitors. And many protected areas, such as national parks, are bordered by human communities of varying sizes. Not all humans have overcome their desire to control nature, so the question arises: Will humans tolerate reintroduced species?

Some species are favored by people. Talk of reintroducing the Eastern bluebird, the salmon, the bald eagle, or the moose, does not cause controversy. On the other hand, talk of wolves and cougars and grizzlies does. Some suspect that hunters might kill wolves released in the Adirondack preserve (McKibben 1988),[8] pointing out the very real problem of human acceptance for restored wilderness areas and species.

Human communities near or within wilderness areas represent very real economic concerns as well. Wolves, although not the bloodthirsty killers legend portrays, might at times seek out livestock. People who established farms in the absence of such predators have legitimate claims which must be included in the policy-making process (McKibben 1988).

Although we ethically cannot ignore human interests, these factors do not seem to constitute a moral obstacle to restoration. Rather, they indicate the need for education, change in attitude, and caution. Environmentalists must educate the general public and specific groups, such as farmers and hunters, before species are reintroduced. While this might delay the reintroduction of some species, it may well lead to greater long-term success.

Economic problems fall into two subsets: costs of restoration and distribution of those costs. Although technology will reduce some costs, restoration is still expensive. How much should we pay for restoration? And should we restore even when the cost is prohibitive?

There seems to be no ready answer, whether economic or moral, to these questions. Berger (1985) has demonstrated that environmental restoration can create jobs and help a region's economy; the same might hold for the more specific case of wilderness restoration. So perhaps benefits might balance costs. However, the choice to begin or to continue the effort to raise and spend the money reflects value priorities, and should not be decided by purely economic cost-benefit analyses:

> The ordinary currencies of economic value are much distorted when one tries to count ecological values with them, for they are poorly equipped to preserve these non-commercial values—those tied to the atmosphere, the oceans, polar ice caps, and ozone layer—which are essential to the health of the ecosystem and thus to human welfare. (Rolston 1986, 77; see also Schumaker 1973)

Economic cost-benefit analyses cannot measure noncommercial values. But in deciding whether to restore wilderness, such values, especially those uniquely associated with wilderness, must be considered (Godfrey-Smith 1979; Rolston 1986). Including such values in an analysis, while not reducing the cost of restoration, will enlarge the picture and clarify what is of importance to us.

As an analogy, we Americans have committed our nation to a space program, despite the cost. This commitment indicates that exploring, searching, and discovering are of great value to us. We could make a similar commitment to restore many open areas to wilderness. First, however, we have to give noneconomic values a larger role in our deliberations.

Some will argue that because these noncommercial values are not quantifiable, they cannot be counted. Some will claim that to count them we must quantify them. Such arguments assume that economic values are primary or that all value is reducible to economic value. But, as with the space program, noneconomic values can and do in fact enter into our decisions.

Like the space program, wilderness restoration will have to compete with such values as human welfare, health, poverty, local cultures, and education. Unlike the space program, however, restoration can contribute to each of those other values, so perhaps it would not fare badly.[9] The cost of restoration presents no reason not to pursue it. In some cases, there may well be good economic reasons not to restore. In other cases, value priorities on use of economic resources may lead to a decision not to restore.

We must also consider the distribution of restoration costs. Some individuals have used their own money, time, and effort to restore some small patch of wilderness (Berger 1985). Perhaps their work is what philosophers call supererogatory: good action which goes beyond obligation. However, in many cases justice seems to demand that despoilers of wilderness ought to pay for its restoration. This is simple restitution—a principle widely recognized, both morally and legally. Unfortunately, it is also widely ignored. Individuals and corporations who have destroyed or degraded wild areas have often not paid for their restoration. This is true of backpackers whose sheer numbers result in degradation of wild areas, as well as of utility companies whose emissions spread acid rain on mountain lakes.

The obstacles to restitution seem to be practical rather than moral: how to affix a just amount for particular damage, and how to force offenders to pay. But courts deal with such problems in other matters all the time. We would also need to decide, in some cases, whether to wait for such restitution before beginning the work. And, if no perpetrator can be found (or none really exists), as perhaps in the case of restoring pieces of natural prairie (Berger 1985), distribution of costs becomes an issue of public social priorities.

Another objection against restoration in general is that the concern for nature shown by those who wish to restore it might encourage environmental destruction. "In some cases, the possibility of restoration at some point in the future may be misinterpreted as a license for further destruction in the present" (Pollock 1988, 52). The forests of the Adirondacks present us with a prime example of this possibility. Given all the abuses that occurred there, some might conclude from the present state of the area that clear-cutting or other practices create only short-term harms which can be corrected later. This is certainly cause for serious concern.

However, definite, consistently enforced penalties for pollution and abuse of resources would nullify this objection. Most arguments against restoration really point to the lack of social commitment to the environment and, perhaps, to the need for a unique environmental ethic. In terms of action, the goals of restoration and its relationship to other environmental issues must be absolutely clear and explicit. Once more, a potential barrier to restoration seems connected to important, practical issues, but not to logical or moral confusion.

Restoration creates another practical problem. Because it raises the question of priorities in a new way, however, it is also a value question. Whether and how much to spend on restoration are questions of social priorities. Similar questions arise for the environmental movement, which itself has limited resources. Some have cautioned against the promotion of restoration, claiming "that it could distract activists who might otherwise be fighting to preserve what little wilderness remains in the world" (Pollock 1988, 52). This objection questions how best to ration the resources of environmentalism and whether to choose preservation or restoration.

Taking the second objection first, keeping a wilderness intact is preferable to having to restore it. The intact area would contain all its natural elements and species while a restored area might not. Nevertheless, it does not follow logically that we focus only on preservation. That would be too strong a conclusion. What does follow is that preservation should take precedence. This seems even more evident given the rapid rate of wilderness destruction occurring today. Restoration should still be practiced wherever and whenever it will not interfere with preservation efforts. That much can be said on conceptual grounds.

But it is difficult, if not impossible, to say whether a particular restoration effort will interfere. For example, restoration might attract those who would not become actively involved in preservation efforts. So it may well be capable of mustering enough independent resources that it is not a drain on other priorities. Besides, restorers might also become strong advocates for preservation, having seen firsthand the difficulties of restoring an ecosystem.

These are only possibilities. The point is to illustrate that the possible consequences of restoration for the preservation movement are not unambiguously negative. But no one can offer an advance solution to the difficult question of whether a specific restoration effort would help or hinder specific preservation efforts.

Finally, what about the potential for confusing restoration with mitigation (Pollock 1988), the required restoration of damaged areas by developers in exchange for damaging a new area? (Whether mitigation itself is morally sound, while certainly debatable, is not the point here.) Possible confusion does not constitute a moral objection to restoration. Even with attempts in the political arena to engender and exploit such confusion, its potential is not grounds for condemning or even delaying restoration.

Restoration, Hubris, and Ignorance

Hubris, extreme pride, is the character flaw of the central figures in classical tragedy and brings about the protagonist's downfall. If a species can exhibit hubris, the human species has undoubtedly done so in its effort to conquer nature, to control the elements, to bend the world to its will. Poets and social critics have pointed to this hubris and have prophesied that it will destroy us, as it does the tragic hero. In this context, the environmental crisis is the beginning of our end.

Is hubris present in the belief that we can restore wilderness? We can restore buildings and cars; they are our creations. But ecosystems invented us. Thinking we could change them at our whim and at no risk, we did so. Now we believe we can fix them. The similarity on this point between the two intentions should shock us into attentiveness.

There is an ambiguity in the rhetoric of restoration enthusiasts. One writes that "restoration is an effort to imitate nature in all its artistry and complexity by taking a degraded system and making it more diverse and productive" (Berger 1985, 5; see also Wolf 1988). This can be seen as either respect and love for nature or as ultimate conceit. In classical tragedy the heroes are unaware of their own hubris. Good intentions are not insurance against our flaws.

Restorationists may just need to be cautioned. Even so, they may not wish to hear this message. Many ecologists vehemently object to having their interventions characterized as "manipulations" of nature. The word has acquired some negative connotations, but I intend it as neutral regarding humans' intentions in intervening in the environment. Any such intervention, however, whether exploring for oil or improving a streambed, manipulates nature and might have unforeseen, destructive consequences. Only the arrogant ignore that possibility.

However, in most restoration projects there seems to be a clear awareness of the fragility, the beauty, and the wholeness of nature (Berger 1985). That is where restoration differs from the domination of nature. The intention of both is to alter some bit of nature. But restoration implies an *attitude* of respect, even humility, toward nature and the awareness that we are a part of nature, not independent of it. Restoration implies a rejection of the attitude of domination humans have so often shown toward nature. But we must couple this new attitude with an awareness of the limitations of our knowledge and power. We cannot pretend to know how to re-create nature.

Our humility will grow as our awareness of our oneness with nature grows. Not our oneness in the spiritual sense, though that too will help, but our oneness in the biological sense. The more we recognize our connection to the biosphere as a plant is connected to the soil, the more we will distance ourselves from hubris and approach humility in our dealings with nature.

A humble question to ask is, should we act in ignorance? We do not know everything there is to know about wilderness ecosystems. We do not know whether a wilderness minus a few native species will prosper or falter. We do not know whether subtle changes will doom reintroduced species, despite our best efforts. We do not know, yet it seems we must act. Wolf argues that

> although tropical forests can slowly reclaim cropland and pasture, from the standpoint of biological diversity, slowly is not enough. Once a tract of forest is reduced to isolated pockets, each of the fragments begins to lose species. . . . Extinctions occur fairly quickly. If the fragments, particularly those amid abandoned land, can instead be rejoined into larger areas quickly enough, at least some extinctions could be prevented. (Wolf 1988, 110)

We must minimize our ignorance by preserving and studying what wilderness is left and by applying that knowledge to restoration. Wolf (1988, 110) says, "Degraded ecosystems cannot be restored to health haphazardly: Restoration requires natural ecosystems as models and seed sources." Once again, the essential link between restoration and preservation emerges. To prevent extinctions, we must both preserve *and* restore. By doing both in a thoughtful and humble manner, we can fulfill part of our duty to nature.

Conclusions

My intention in presenting the major arguments for and against wilderness restoration has been to determine whether we have a moral duty to restore wilderness areas and whether there are any moral or normative considerations which make such restoration efforts objectionable. None of the latter have been found. On the other hand, the arguments favoring restoration make a case for preservation—not surprising, given the close connections between the two. However, while not surprising, this conclusion is significant. It points to the need to place preservation higher on our list of environmental priorities. This is both a practical imperative (we cannot restore without preserved examples to learn from) and a normative one (ecosystems are valuable to begin with and thus should be preserved). The objections to restoration fall into two categories: those which point to practical problems connected with implementing restoration efforts on a large scale and those which dictate caution and concern for particular aspects of restoration (e.g., fairness to species to be reintroduced into an area).

As for the prorestoration considerations, they establish a duty to restore wilderness areas when such efforts do not interfere with preservation attempts. Perhaps the strongest argument for restoration is the one based on species extinctions, though that too indicates the value of preserving wilderness.

The Adirondack Park has within it areas which are preserved, and many acres have been recovered from near destruction. We can learn a great deal from such efforts. Not the least lesson points to how much harm we have done ourselves in destroying wilderness and how, in healing the earth, we heal ourselves.

Notes

Author's note: Comments from John Spissinger, Mike Andolina, and Bonnie Steinbeck helped me to improve this paper. Any remaining shortcomings are my responsibility.

1. It is tempting to call this the active/passive restoration distinction analogous to the active/passive euthanasia distinction. However, I think that in the present context such terminology would be misleading for two reasons: First, restoration is analogous to the healing process, not to euthanasia; second, in euthanasia we're dealing with humans who have choices, whereas in restoration we're dealing with ecosystems, which do not. This is *not* to say, however, that ecosystems cannot play a role in our moral deliberations. I contend that they should.

2. John Spissinger brought this criticism of the historical restoration argument to my attention.

3. De George (1979, 95) argues that although we can say that future generations have rights, "we avoid more problems if we maintain that, *properly speaking*, future generations do not presently have rights, than if we say they do" (emphasis added).

4. In his essay, "Are Values in Nature Subjective or Objective?" Rolston (1986) provides a detailed critique of the notion that value is solely related to humans and their experiences. Ernest Partridge (1986) criticizes Rolston's view.

5. Although he does not label it as restoration, Rolston (1988) argues in much the way I interpret him here in justifying the re-establishment of a rare species of trout in California. Paul Taylor (1986) argues that humans can incur duties of restitution to organisms, species, populations, or biotic communities. Restoration would be one form of restitution.

6. Joel Feinberg (1974) argues that we have no duties to species because species *per se* have no interests (whereas individual sentient animals do have interests, so we have duties toward them). Alastair Gunn (1984) argues that by threatening species we threaten ecosystems (and by threatening ecosystems we threaten species!), which are good in themselves; but he refuses to say species *as such* are valuable. Rolston agrees that ecosystems are intrinsically good, but extends this view to species as well.

7. J. Baird Callicott (1980) has argued that animal liberation cannot serve as the basis of an environmental ethic because of the former's focus on the suffering of individual animals. An environmental ethic must value species, forests, lakes, ecosystems, in which some individual animals (e.g., prey) in fact suffer. I believe that, whatever the divergence between animal libera-

tionists and environmental ethics theorists, they might agree on the point I am making here: For the liberationists, it would apply to each individual animal released; for the environmental ethics theorist, it would apply to the species.

8. This fear was expressed by a NYS Department of Environmental Conservation official. According to McKibben (1988), several hunting organizations and farmers are opposed to the reintroduction of wolves. It may well be that overall these groups are divided on the issue.

9. Those connections are already being made explicit. Karen J. Warren (1987, 1990) has shown that the domination of women and the domination of nature are connected historically, symbolically, and conceptually. And Kazis and Grossman (1982) show that the labor and environmental movements are allies, not enemies.

References

Berger, J. J. Restoring the earth: How Americans are working to renew our damaged environment. New York: Knopf; 1985.

Cairns, J. Jr., editor. The recovery process in damaged ecosystems. Ann Arbor, MI: Ann Arbor Science Publishers; 1980.

Callahan, D. What obligations do we have to future generations? The American Ecclesiastical Review 164:4; 1971.

Callicott, J. B. Animal liberation: A triangular affair. Environmental Ethics 2:311–338; 1980.

De George, R. T. The environment, rights and future generations. In: Goodpaster, K. E.; Sayre, K. M., eds. Ethics and problems of the 21st century. Notre Dame; IN: University of Notre Dame; 1979:93–105.

Feinberg, J. The rights of animals and unborn generations. In: Blackstone, W., ed. Philosophy and environmental crisis. Athens, GA: University of Georgia Press; 1974:43–68.

Fox, M. A. Animal liberation: A critique. Ethics 88:106–118; 1981.

Godfrey-Smith, W. The value of wilderness. Environmental Ethics 1:309–319; 1979.

Gunn, A. Preserving rare species. In: Regan, T., ed. Earthbound: New introductory essays in environmental ethics. Philadelphia: Temple University Press; 1984:289–335.

Kazis, R.; Grossman, R. L. Fear at work: Job blackmail, labor and the environment. New York: Pilgrim; 1982.

McKibben, B. The return of a native. Adirondack Life 1988:19; 1988.

Nash, R. Wilderness and the American mind. 3d ed. New Haven, CT: Yale University Press; 1982.

New York State Department of Environmental Conservation and U.S. Fish and Wildlife Service. Draft environmental impact statement: Use of lampricides in a temporary program of sea lamprey control in Lake Champlain. New York: Ray Brook; 1987.

Partridge, E., editor. Responsibilities to future generations. Buffalo, NY: Prometheus; 1981.

————. Values in nature: Is anybody there? Philosophical Inquiry 8:96–110; 1986.

Pollock, S. A time to mend. Sierra (Sept.-Oct.):51–55; 1988.

Postel, S.; Heise, L. Reforesting the earth. In: Brown, L. R.; Chandler, W. U.; Durning, A.; Flavin, C.; Heise, L.; Jacobson, J.; Postel, S.; Pollock Shea, C.; Starke, L.; Wolf, E. C., eds. State of the world: 1988. New York: Norton; 1988:83–100.

Rolston, H., III. Philosophy gone wild: Essays in environmental ethics. Buffalo, NY: Prometheus; 1986.

————. Environmental ethics: Duties to and values in the natural world. Philadelphia: Temple University Press; 1988.

Schumaker, E. F. Small is beautiful: Economics as if people mattered. New York: Harper and Row; 1973.

Singer, P. Animal liberation: A new ethics for our treatment of animals. New York: Avon; 1975.

Taylor, P. Respect for nature: A theory of environmental ethics. Princeton, NJ: Princeton University Press; 1986.

Turk, J. Introduction to environmental studies. 2d ed. Philadelphia, PA: Saunders; 1985.

Warren, K. J. Feminism and ecology: making connections. Environmental Ethics 9:3–20; 1987.

————. The power and promise of ecological feminism. Environmental Ethics 12:125–146; 1990.

Wolf, E. C. Avoiding a mass extinction of species. In: Brown, L. R.; Chandler, W. U.; Durning, A.; Flavin, C.; Heise, L.; Jacobson, J.; Postel, S.; Pollock Shea, C.; Starke, L.; Wolf, E. C., eds. State of the world: 1988. New York: Norton; 1988:101–117.

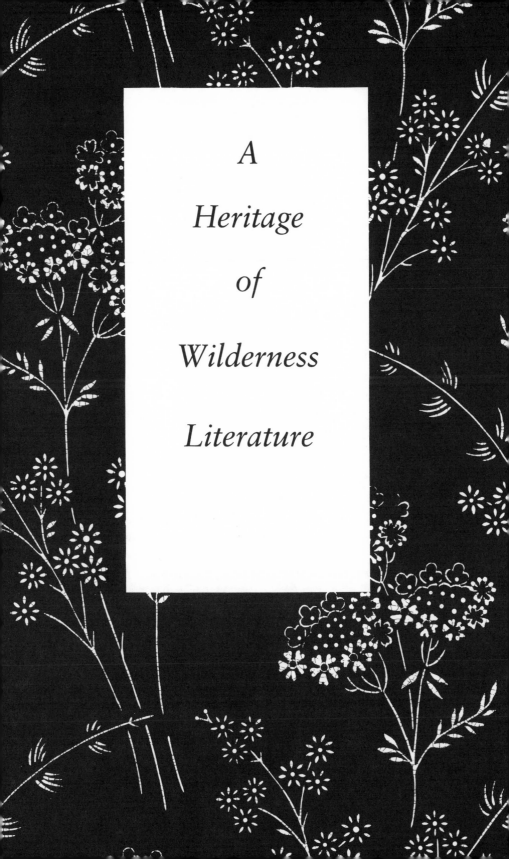

A

Heritage

of

Wilderness

Literature

Thoreau's Language of Ecology

WILLIAM C. JOHNSON, JR.

S C H O L A R S have gradually recognized that Thoreau's journal represents not a jumble of randomly collected facts, but a
disciplined testament recording the growth of a mind in nature,
and nature in a mind (*cf.* Cameron 1985). It is, in his own words,
"a book of the seasons, each page of which should be written in its
own season and out-of-doors, or in its own locality wherever it may be"
(11 June 1851).[1] The journals weave together a manifold of meanings:
field-observations, excerpts from and commentaries on his voluminous
reading (especially in literature, botany, and natural history), and ideas
as they come to him either spontaneously and upon later and deeper reflection. This examination shows how the language of Thoreau's journal
becomes a valuable instrument for understanding the natural world.

In the summer of 1845, Thoreau (1984, 1:205–206) complained that
even though he faithfully records inspirations "in such order as they
come" and works them into lectures, and the lectures in turn into essays,
they nonetheless remain like statues on pedestals and do not hold hands.
Early in his writing career, Thoreau recognized that lectures and essays
could not provide the "horizontal connection" he at first was seeking.
Scholars have seized upon this entry to remind us that *A Week on the
Concord and Merrimack Rivers* and *Walden* exemplify the organic unity
Thoreau felt lacking in his earlier work.[2] But the entry also reminds

us that, from the beginning of his career, Thoreau used the journals as a unique genre, a document valuable precisely because it allowed him to record observations "in such order as they come." As early as January 30, 1841, he wrote, "Of all strange and unaccountable—things this journalizing is the strangest, it will allow nothing to be predicated of it; its good is not good, nor its bad bad" (Thoreau 1981, 1:237). It is, he continued, like the "homemade stuffs" on his counter that will, in time, become the wealth of India; a "confused heap" like festoons of "dried apple or pumpkin" that will become "diamonds or pearls."

Over the years, Thoreau's painstaking *Journal*, full of inspirations, meticulous and loving field-observations, and random vitality, reveals a linguistic analogue of life as he actually lived it, a life that comes ever more to be a surrogate of the life of nature. On January 27, 1852, he wrote:

> I do not know but (that) thoughts written down thus in a journal might be printed in the same form with greater advantage than if the related ones were brought together into separate essays. They are now allied to life, and are seen by the reader not to be farfetched. It is more simple, less artful. I feel that in the other case I should have no proper frame for my sketches. Mere facts and names and dates communicate more than we suspect. Whether the flower looks better in the nosegay than in the meadow where it grew and we had to wet our feet to get it! Is the scholastic air any advantage? (Thoreau 1962, 334)

And the next day, deepening his response to the previous entry, he commented, "Perhaps I can never find so good a setting for my thoughts as I shall thus have taken them out of. The crystal never sparkles more brightly than in the cavern." In a word, Thoreau realized that the *Journal* reflects a primal relation to the natural world, a relation meaningful apart from institutional or ideological strictures superimposed on nature. It becomes an experiment seeking to align language not to conventional themes, such as friendship (*A Week on the Concord and Merrimack Rivers*) or the conflict between self-realization and social institutions (*Walden*), but to nature itself, as a living presence, separate from, yet intrinsic to, human culture.

In this immediacy, the *Journal* is a markedly ecological document. It reminds us that ecology, as "the study of the relationships between organisms and their environment" (Keeton 1970, 709), must also include the human observer.[3] The long-standing habit of excerpting vignettes or set pieces from the *Journal* (analogous to studying an organism apart from its ecological niche) has impeded recognition of the uniqueness of Thoreau's text, the "generic strangeness" (Cameron 1985, 47) that reveals the growth of his mind and sensibility.[4] The actual *Journal* reveals the recursive deepening of Thoreau's mental, emotional, and physical perambulations through his Concord "neighborhood." Increasingly, after 1850, the basis, or trigger, of journal entries is his observation of the Concord environment—its landforms, plants, animals, fish, birds, water, and light. And his language repeatedly addresses "facts" out of an awareness that the observer himself is a living part of the field. As early as February 8, 1841, he wrote, "As time is measured by the lapse of ideas—we may grow of our own force—as the mussel adds new circles to its shell—My thoughts secrete the lime." And in the next paragraph, "My journal is that of me which would spill over and run to waste—gleanings from the field which in action I reap. I must not live for it, but in it for the gods—."

Just what this growth and reaping mean becomes increasingly clear in his entries during the late 1840s into the 1850s, when Thoreau's reading in natural history and his accumulative observations of seasonal flora and fauna move him deeper into a polar interplay of natural fact and consciousness. On February 20, 1857, four years before his death, he wrote:

What is the relation between a bird and the ear that appreciates its melody, to whom, perchance, it is more charming and significant than to any else? Certainly they are intimately related, and the one was made for the other. It is a natural fact. If I were to discover that a certain kind of stone by the pond-shore was affected, say partially disintegrated, by a particular natural sound, as of a bird or insect, I see that one could not be completely described without describing the other. I am that rock by the pond-side. (Thoreau 1962, 1121)

This is the Hindu *tat twam asi* ("that art thou"), with an empirical twist. The language implies a psychic interpenetration of fact and spirit. Nature and consciousness are truly interdependent; the bird's song and the human ear are convergent in the act of knowing. "It is a natural fact."

And there is the further suggestion that to know is to become what is known, "I am that rock by the pond-side." Language is the instrument joining nature and mind in one ecology. Such observations make clearer Thoreau's many statements on the virtues of "natural" expression, his praise of Saxon word-roots, his conversational diction, and his frequent analogies between plant-growth and linguistic expression.

Thoreau's method of entry-taking, furthermore, reflects his awareness of growth as conscious process, a reflective, creative accretion Sattelmeyer and Hocks (1985) have called the "germ theory" of composition (vol. 2, 453). Late in his career, Thoreau describes his method in this way:

> I would fain make two reports in my Journal, first the incidents and observations of to-day; and by to-morrow I review the same and record what was omitted before, which will often be the most significant and poetic part. I do not know at first what it is that charms me. The men and things of to-day are wont to lie fairer and truer in to-morrow's memory. I saw quail-tracks some two months ago, much like smaller partridge-tracks. (Thoreau 1962, 1129; March 27, 1857)

Thoreau habitually uses a "double-entry" approach, deepening his original observations through further commentary. This particular entry suggests that writing and field observation can act as mutual triggers, the general comment on "tomorrow's memory" apparently leading back to an actual memory of quail tracks.

Thoreau's specific practices in composing have recently been clarified by editors of the Princeton edition of the *Journal*, who have scrutinized the original manuscripts (R. Sattelmeyer and T. Blanding [see Thoreau 1981, 1984]). Sattelmeyer and Hocks (1985) point out that Thoreau left large blank spaces above and below each entry. He later expanded these entries, adding further observations through retrospective reflection. This suggests that he viewed the original entries as germinal seeds, to be cultivated and fed by thought. This habit became more systematic during 1849–1851, when Thoreau worked on the later phases of *Walden* and immersed himself in natural history, such as Darwin's *The Voyage of the H. M. S. Beagle,* and botanical and zoological writings, such as Bigelow's *Medical Botany* and Agassiz and Gould's *Principles of Zoology* (Richardson 1986). William Rossi explains:

As before in the experiment whose results fill 'Long Book,' Thoreau developed his 'germs.' Now, however, they were gathered daily on his walks, and instead of going immediately into the Journal were jotted shorthand in a notebook and then expanded in the Journal after a short period of time. (Rossi 1986, 36)

This method of composing admits that observation is not merely empirical and specific, but ideal and inclusive. In Thoreau's words, "most that is first written on any subject is a mere groping after it, mere rubblestone and foundation. It is only when many observations of different periods have been brought together that he begins to grasp his subject and can make one pertinent and just observation" (February 3, 1859). The "seed" of the initial observation requires the attention of the whole being, and its linguistic recreation occurs most effectively through a progressive deepening, a return to the observation, through language, over time, in the reserves of memory. Rossi (1986) points out that Thoreau took care starting around 1850 to preserve the journal leaves intact in a volume, instead of removing them for other projects as he had before the *Walden* period. He came to value the journal itself as a primary form of expression, a verbal embodiment of his growth in nature and nature's in him.[5]

The journal increasingly reflects a polar interplay of fact and spirit, nature and consciousness, each taking the other on through language. And language increasingly reflects what Thoreau observed in walking the Concord woods and fields. On March 1, 1860, an entry about walking in the rain ends with the following summary:

I have thoughts, as I walk, on some subject that is running in my head, but all their pertinence seems gone before I can get home to set them down. The most valuable thoughts which I entertain are anything but what *I* thought. Nature abhors a vacuum, and if I can only walk with sufficient carelessness I am sure to be filled. (Thoreau 1962, 1595)

Such entries may reveal that "phenomena are disassociated from human significance" (Cameron 1985, 75), that man is a witness and beholder, not an explicator of nature. However, while Thoreau may have come to eschew conventional genres in his quest for recording nature's plenty, he nonetheless attempted to extend and deepen, to humanize in the direction of the natural and vice-versa, to let nature's facts permeate and color his thinking, so that thought and phenomena become mutually informing.

Given the ability of human consciousness both to lose and find itself, such interplay is at once engaging and disengaging, immersed and detached. What makes it possible at all, however, is meeting facts in their *own* domain, and on their own terms, and only afterward attending to their possible significance. On March 15, 1860, Thoreau remarked that he had never seen a hen-hawk accurately represented in any book:

> I have no idea that one can get as correct an idea of the form and color of the under sides of a hen-hawk's wings by spreading those of a dead specimen in his study as by looking up at a free and living hawk soaring above him in the fields. The penalty for obtaining a petty knowledge thus dishonestly is that it is less interesting to men generally, as it is less significant. (Thoreau 1962, 1601)

For Thoreau, "significance" transcends empirical imprisonment of the object. Instead, it entails direct experience of nature itself, untrammeled by the preemptory activities of scientific reason.

It is, likewise, no longer possible to accuse Thoreau of obscurantism or naivete about the science of his own day. As Rossi (1986) and Richardson (1986) have shown, he was deeply conversant with science and increasingly critical of its positivistic thrust. Early in his career, in "The Natural History of Massachusetts" (published in *The Dial*, July, 1842) he wrote:

> The true man of science will know nature better by his finer organization; he will smell, taste, see, hear, feel, better than other men. His will be a deeper and finer experience. We do not learn by inference and deduction and the application of mathematics to philosophy, but by direct intercourse and sympathy. It is with science as with ethics,—we cannot know truth by contrivance and method; the Baconian is as false as any other, and with all the helps of machinery and the arts, the most scientific will still be the healthiest and friendliest man, and possess a more perfect Indian wisdom. (Thoreau 1980, 29)

Thoreau deliberately chose a path that would join, not sever, scientific and poetic truth. What he wrote on March 5, 1853, holds true for the last decade of his career. "The fact is I am a mystic, a transcendentalist, and a natural philosopher to boot."

Thoreau's reading and excerpting of Coleridge's *Theory of Life* in 1848, moreover, gave him a theory of nature that combines poetry and empiricism. Trevor Levere's (1981) description of Coleridge's method applies equally well to Thoreau. "When Coleridge went to the ant, it was not only with a naturalist's eye, but also with a view to extending his comprehension of the role of powers in mind and nature, the correspondences among them, the teleology directing them, and the language embracing and incorporating them" (Levere 1981, 213). Like Thoreau, Coleridge counters scientific positivism with holistic awareness. He holds, for example, that life is not immediately dependent on physical organization but grows from deeper principles. He seeks to clarify these principles and inner agencies of life as *anterior* to organization (Sattelmeyer and Hocks 1985), identifying evolving patterns of polarization and differentiation. Coleridge describes the ascending scale of creation from mineral to vegetable to animal to man, as a successive manifestation of the *inner* organic aptitudes of irritability, reproduction, and sensibility. Each level in the scale of creation synthesizes and reexpresses what it leaves behind. The bird, for example, realizes potentialities of both fishes and winged insects. Man takes his place as a unique yet representative being in the larger scale. In Coleridge's words, from a passage Thoreau copied into his journal: "In Man the centripetal and individualizing tendency of all Nature is itself concentrated and individualized—he is a revelation of Nature!" (Coleridge 1951).

Mere summary cannot capture the intricacies of Coleridge's argument, which shares important affinities with Thoreau's, especially in relation to growth out of generative polar conflict. In the *Journal*, language alternately engages and detaches itself from phenomena, allowing them to *be* what they are, and allowing the observer both sensuous and reflective responses to them. Thoreau's observations are meticulously detailed, but he mingles them with metaphoric analogies, and a richly melodic prose, to create parallel alternatives, complements to scientific discourse, a virtual biology of consciousness that allows the whole of nature, not merely its metrical side, to be known. Thoreau's stylistic shifts, sudden disruptions, penchant for myth, anecdote, reverie, precise description, and measurement, in sum, reflect the richness of nature itself, far beyond any consistent, one-dimensional cataloguing of nature.

Such also fairly describes his journal-keeping. But the best way to demonstrate Thoreau's method is to observe him, as it were, in action. Thoreau's journeys around Concord were both cumulative and recursive. Returning to the same haunts during the same season in successive years, he deepened his empirical and sympathetic observations through memory, comparative descriptions, and judgment. Whether snowflakes, ice, crows, tortoises, mosses, oaks, pines, crickets, fishes, or numerous other flora and fauna, Thoreau approached them with both loving reverence and the patience of a professional naturalist. His study of lichens serves to illustrate the nature of this quest.[6]

Between November 8, 1850, and December 24, 1859, Thoreau made some fifty-eight references to lichens, the bulk of them field observations. His study of lichens worked in two directions, drawing us inward to their botanical uniqueness and leading us outward to their philosophical and spiritual significance. Thoreau sought them out in their own ecological niches. They were, for him, key signs that winter lived on into spring, for many of his observations occurred in early March and were associated with signs of spring. On March 7, 1853, for example, he remarked that lichens were fresh and brilliant as he collected them, but shriveled up by the time he got home, a thought that led him to reflect that most men are "natural mummies," lifelike, but dried up, like the body of a dead snipe he discovered. Perhaps Thoreau took up the study of lichens because Channing dropped it, because to Channing lichens were "so thin," had so little of man in them (January 11, 1852) (see also Channing 1902).

About a month after recording Channing's remarks, Thoreau speculated that the apothecium or cupped spore-casings of lichens appeared to be all fruit (February 16–17, 1852), and then speculated further as to just what kind of botanist he himself was. He had been reading Linnaeus's *Philosophia Botanica*, and he commented, "By the artificial system we learn the names of plants, by the natural relations to one another; but still it remains to learn their relation to man. The poet does more for us in this department." And he sees himself in Linnaeus's classification of types of botanists as "one of those who describes various plants, whose observations are not (strictly) in accord with botanical science, but one who exclaims the praises of plants, like a poet" (author's translation).

On March 23, 1853, lichens were the catalyst for one of Thoreau's well-known critical statements on the nature of scientific observation:

Man cannot afford to be a naturalist, to look at Nature directly, but only with the side of his eye. He must look through and beyond her. To look at her is fatal as to look at the head of Medusa. It turns the man of science to stone. I feel that I am dissipated by so many observations. I should be the magnet in the midst of all this dust and filings. I knock the back of my hand against a rock, and as I smooth back the skin, I find myself prepared to study lichens there. I look upon man but as a fungus. I have almost a slight, dry headache as the result of all this observing. How to observe is how to behave. O for a Little Lethe! To crown all, lichens, which are so thin, are described in the *dry* state, as they are most commonly, not most truly, seen. Truly, they are *dryly* described. (Thoreau 1962, 539)

Unfortunately, this statement has often been used to argue that Thoreau's imaginative powers were drying up in his "late" years, as a result of too much scientific observation. It seems clear, however, that while Thoreau certainly felt the dryness he described, he was at the same time exploring its very origin and its remedy, indeed, offering an alternative way of seeing. He said he was "dissipated" by encountering nature only through the "dust and filings" of dead specimens, clinically described. Thoreau studied lichens *in situ*, as an appropriate metaphorical equivalent, or better, homological parallel, of his own bodily experience. "I knock the back of my hand against a rock, and as I smooth back the skin, I find myself prepared to study lichens there." The skin of the hand is analogous to lichen growing on rock. The entry reveals not that Thoreau has "dried up," but that he is still imaginatively alive, still capable of saying, "I look upon man as a fungus." We understand what we see by allowing nature to work in and through us. And clearly, lichens are best described, not in their dry, but in their green, living states, as Thoreau often tells us (e.g., March 3, 1857).

Thoreau's long and loving study of lichens culminated in a journal entry for February 7, 1859, which deserves to be quoted in full:

Going along the Nut Meadow or Jimmy Miles road, when I see the sulphur lichens on the rails brightening with the moisture I feel like studying them again as a relisher or tonic, to make life go down and digest well, as we use pepper and vinegar and salads. They are a sort of winter greens which we gather and assimilate with our eyes. That's the true use of the study of lichens. I expect that the lichenist will have the keenest relish for Nature in her every-day mood and dress. He will have the appetite of the worm that never dies, of the grub. To study lichens is to get a taste of earth and health, to go gnawing the rails and rocks. This product of the bark is the essence of all times. The lichenist extracts nutriment from the very crust of the earth. A taste for this study is an evidence of titanic health, a sane earthiness. It makes not so much blood as soil of life. It fits a man to deal with the barrenest and rockiest experience. A little moisture, a fog, or rain, or melted snow makes his wilderness to blossom like the rose. As some strong animal appetites, not satisfied with starch and muscle and fat, are fair to eat that which eats and digests,— the contents of the crop and the stomach and entrails themselves,—so the lichenist loves the tripe of the rock,—that which eats and digests the rocks. He eats the eater. 'Eat-all' may be his name. A lichenist fats where others starve. His provender never fails. What is the barrenest waste to him, the barest rocks? A rail is the sleekest and fattest of coursers for him. He picks anew the ones which have been picked a generation since, for when their marrow is gone they are clothed with new flesh for him. What diet drink can be compared with a tea or soup made of the very crust of the earth? There is no such collyrium or salve for sore eyes as these brightening lichens in a moist day. Go and bathe and screen your eyes with them in the softened light of the woods. (Thoreau 1962, 1424)

The passage is representative of Thoreau's ecology throughout the 1850s. Its earthy life and metaphorical richness harken back to the sandbank passage in *Walden*'s "Spring" ("There is nothing inorganic"). Yet there is decidedly less rhetorical flourish and intellectualizing here. Indeed, Thoreau could have developed lichens as another example of the archetypal Goethean leaf, "the principle of all the operations of Nature" (*Walden* XVII, 9), but instead he chose to show, simply, that lichens are what "make life go down well." The language nourishes symbiosis between the mind and lichens and is a prime example of Thoreau's belief that to know is to become. Earth and lichens are gastronomic partners in taste and eating ("relish," "tonic," "greens," "appetite," "gnawing," "crust," "tripe"). Lichens represent the power of universal assimilation since they eat and thus become their host, just as imaginative language eats the world. The phrase "titanic health" is at once ecology and myth. The Titans were the children of Uranus and Gaea, Sky and Earth, hence studying lichens evidences a "sane earthiness," a "sound" or "healthy" (*sanus*) relationship with the earth deeper than blood, that "makes not so much blood as soil of life." Thoreau would remind us that a proper biology (*bios*, 'life') is inseparable from a proper physics (*phusis* from *phuein*, 'to bring forth, make grow'). Nature's history is also biography. Thoreau ended with an analogy between lichens and human vision. Lichens are a "salve for sore eyes" and *Walden*'s injunction to bathe in the pond and read the *Vedas* now becomes explicitly ecological. "Go and bathe and screen your eyes with them in the softened light of the woods."

If this sounds far-fetched to contemporary scientific ears, a romantic hankering after the German *Natürphilosophie*, a closer look at developments in what is now being called the "new biology" indicates that Thoreau anticipated important aspects of current scientific thinking. In E. S. Russel's words,

> biology occupies a unique and privileged position among the sciences in that its object, the living organism, is known to us not only objectively through sensory perception, but also in one case directly, as the subject of immediate experience. It is therefore possible, in this special case of one's own personal life, to take an inside view of a living organism. (Russel 1930, 138)

Once again, the scientist himself can become one of the instruments of search and research, and the language of percept and image as important as that of concept and definition. The scientist and poet are but complementary observers. And this, of course, recalls that Goethe long ago urged us, in Amrine's words, "to regain the qualitative dimension lost through mathematized reduction by allowing it to be shaped and guided by the phenomena themselves" (Amrine 1983, 23). And this means phenomena observed not only through metrical measurement, but through the human sense organs and emotions. Here Thoreau's ecology is exemplary.

Thoreau shares central concerns with perhaps the most trenchant critic of contemporary scientific positivism, Owen Barfield, who, in elucidating Coleridge's critique of science, explains the premise on which much of today's science is still largely based. Barfield (1973), like Coleridge before him, points out that conventional science is based upon a speculative assumption that has not been and cannot be proved—the absolute dichotomy between mind and matter, subjective and objective, observer and observed. If, on the other hand, physical processes can be distinguished but not separated from mental processes, but the two rather stand dynamically conjoined in a synthesis of opposing energies, or polarities, then Thoreau, in his own scientific endeavors, provides us with an example of how to go about founding a new biology. But such a founding will be extraordinarily difficult, especially for biologists themselves, for it requires full acceptance of the still radical notion that "natural science is self-knowledge" (Barfield 1973, 140). And since ecology is still a young science, it perhaps will become the testing ground for such a vital idea. As Barfield notes, "If polarity is in truth a universal law of nature and of spirit, and is at the same time the basic law of the relation *between* nature and spirit, then it may well need to be recognized, perhaps more than anywhere else, as the 'initiative idea' of ecology" (Barfield 1973, 142).

Like Thoreau, Barfield urges upon us the full range of our experience as knowers, carers, and doers. Insofar as it allows the human observer to shepherd the process of knowing and hence to care about the applications of what is known, ecology must become the foundation of all science. We need not abandon empirical precision, only infuse it with love, remembering, with Heidegger (1977, 213), that "Language is the house of Being," that nature's prime significance is its presence as the preconditioning signified, the presence of Being itself, insofar as it is at once revealed and concealed in nature's appearances. An ecological language admits that what *is* has primal significance over what it means. This premise, combined with Coleridge's principle of polarity, provides the basis of Thoreau's language of ecology. Let us hope it may come to stand as a foundational principle in the science of the twenty-first century.

Notes

1. *Journal* quotations are from the new Princeton edition (1981, 1984) through 1848 (cited by volume number, date, and page number). References after 1848 are to the Torrey & Allen edition of 1906 (1962) (cited by date and page number only).

2. See J. Lyndon Shanley, *The Making of Walden* (Chicago: University of Chicago Press, 1957), p. 75.

3. Thoreau may actually have coined the term "ecology." The OED at-

tributes the honor to the German naturalist Ernst Heinrick Haeckel in the year 1879. But in a letter to George Thatcher of Bangor, Maine, dated January 1, 1858, Thoreau writes: "Mr. Hoar is in Concord, attending to Botany, Ecology, & with a view to make his future residence in foreign parts more truly profitable to him." See Paul H. Oesher (1959). [*Editor's note:* It is uncertain if Thoreau actually wrote the word "Ecology" here; it may have been a similar word such as "Economy."]

4. For a survey of early manuscript history of the journal, see the Princeton edition, volume 1, pp. 578–590.

5. Channing's (1902) observations on Thoreau's walking habits provide a firsthand glimpse of Concord's first full-fledged ecologist at work:

In these walks, two things [Thoreau] must have from his tailor: his clothes must fit, and the pockets, especially, must be made with reference to his out-door pursuits. They must accommodate his note-book and spy-glass; and so their width and depth was regulated by the size of the note-book. It was a cover for some folded papers, on which he took his out-of-door notes; and this was never omitted, rain or shine. It was his invariable companion, and he acquired great skill in conveying by a few lines or strokes a long story, which in his written Journal might occupy pages. Abroad, he used the pencil, writing but a few moments at a time, during the walk; but into the note-book must go all measurements with the foot-rule which he always carried, or the survey's tape that he often had with him. Also all observations with his spy-glass (another invariable companion for years), all conditions of plants, spring, summer, and fall, the depth of snows, the strangeness of the skies,—all went down in this note-book. To his memory he never trusted for a fact, but to the page and the pencil, and the abstract in the pocket, not the Journal. I have seen bits of this note-book, but never recognized any word in it; and I have read its expansion in the Journal, in many pages, of that which occupied him but five minutes to write in the field."

The editors of the Princeton edition of the *Journal* point out that in a journal entry for November 9, 1851, Thoreau described Channing's unsuccessful attempt to keep such a notebook:

In our walks C. takes out his note-book sometimes and tries to write as I do, but all in vain. He soon puts it up again, or contents himself with scrawling some sketch of the landscape. Observing me still scribbling, he will say that he confines himself to the ideal, purely ideal remarks; he leaves the facts to me.

It seems clear that Thoreau's interest in facts is not of a generalized, Wordsworthian nature, but, as with Coleridge, a poetry of facts sharply, but also fully, realized.

6. Angelo (1983) has provided a thorough account of Thoreau's achievements as a botanist. He notes that by 1857 Thoreau was "one of the more competent amateur botanists in Massachusetts" (20) and that in July, 1858, he ascended Mount Washington in New Hampshire "and prepared the most detailed list of plants by zones that had ever been made for this site, one not to be surpassed until the twentieth century" (21). Angelo observes that in Thoreau's time lichens, mosses, and fungi "resisted study owing to the absence of good regional manuals" and that, as a result, Thoreau never came close to acquiring expertise comparable to what he achieved with vascular plants (22). While his technical grasp of lichens may have been slight, he knew their morphological types and appreciated their place in nature. Indeed, Thoreau's lack of sheerly technical expertise in the case of lichens makes his study of them all the more interesting as an illustration of his experiments with a "literary ecology."

References

Amrine, F. Goethe's science in the twentieth century. Towards 2(4):20–23; 1983.

Angelo, R. Thoreau as botanist. The Thoreau Quarterly 15:15–31; 1983.

Barfield, O. What Coleridge thought. Middletown, CT: Wesleyan University Press; 1973.

Cameron, S. Writing nature: Henry Thoreau's journal. New York: Oxford University Press; 1985.

Channing, W. E. III. (Entry.) In: Sandborn, F. B., ed. Thoreau: The poet-naturalist. Boston: Charles E. Goodspeed; 1902:65–66.

Coleridge, S. T. Hints toward a more comprehensive theory of life. In: Stauffer, D. A., ed. Selected poetry and prose of Coleridge. New York: Modern Library; 1951:558–606.

Heidegger, M. Letter on humanism. In: Krell, D. F., ed. Martin Heidegger: Basic Writings. New York: Harper & Row; 1977:213.

Keeton, W. T. Biological science. New York: Norton; 1970.

Levere, T. H. Poetry realized in nature: Samuel Taylor Coleridge and early nineteenth century science. Cambridge, England: Cambridge University Press; 1981.

Oehser, P. The word 'ecology'. Science 129 (335):992; 1959.

Richardson, R. D. Jr. Henry Thoreau: A life of the mind. Berkeley: University of California Press; 1986.

Rossi, W. J. Laboratory of the artist: Henry Thoreau's literary and scientific use of the journal, 1848–1854. Ph.D. dissertation. St. Paul: University of Minnesota; 1986.

Russel, E. S. The interpretation of development and heredity. A study in biological method. Oxford, England: Oxford University Press; 1930.

Sattelmeyer, R.; Hocks. R. A. Thoreau and Coleridge's theory of life. Studies in the American Renaissance 1985: 269–284.

Shanley, J. L. The making of Walden. Chicago: University of Chicago Press; 1957.

Thoreau, H. D. The correspondence of Henry David Thoreau. In: Harding, W.; Bode, C., eds. New York: New York University Press; 1958.

———. The journals of Henry David Thoreau. Torrey B.; Allen, F. H., eds. 14 vol. Boston: Houghton Mifflin; 1906. Rpt. in 2 vol. New York: Dover; 1962.

———. Walden. Shanley, J. L., ed. Princeton, NJ: Princeton University Press; 1971.

———. The Natural History Essays. Sattelmeyer, R., ed. Salt Lake City: Peregrine Smith; 1980.

———. Journal 1:1837–1844. Wetherell, E.; Howarth, W. L.; Sattelmeyer R.; Blanding, T., eds. Princeton, NJ: Princeton University Press; 1981.

———. Journal 2:1842–1848. Sattelmeyer, R., ed. Princeton, NJ: Princeton University Press; 1984.

Wallace Stegner's Version of Pastoral or "No Eden Valid Without Serpent"

RUSSELL BURROWS

A GREAT deal of the vitality inWallace Stegner's fiction derives from his use of pastoralism. In some part, Stegner's pastoral impulse is traditional; he draws on the sensibility that so interested Wordsworth and his critic, W. W. Greg (1906, 10), both of whom spoke of the human heart recoiling from the "world [that] is too much with us." And beyond this theme, Stegner's settings are often pastoral. Some settings in his novels suggest Gail Finney's (1984, 12) definition of the pastoral garden, where events run "counter to the progressive tide of time, social norms, [and] urban civilization."

But Stegner also builds on tradition. His version of pastoral is, as Renato Poggioli says of modern pastorals generally, "self-conscious" and "inverted," so much so that one might say Stegner's gardens offer "a bucolic aspiration only to deny it" (Poggioli 1975, 33–34). In this sense, his gardens resemble one that Steinbeck created in "The White Quail," in which a woman tries to keep herself unspotted from the world but succeeds only in perverting nature and making herself "untouchable" (1938, 30). That her unhappy husband cannot rescue her from this false ideal helps to place the story in a category of pastoral that Andrew Ettin (1984) identifies: those stories that not only comment on contemporary events, but criticize them as well. Prominent in this strain of criticism is William Empson, whose chapter "Proletarian Literature" in *Some Versions of Pastoral* (1935), explains how the convention has come to speak for working classes. This line of criticism also includes Leo Marx, who demonstrates that many narratives, among them Vergil's *Ecologues*, have "political overtones" (1964, 20).

Leo Marx's *The Machine in the Garden* (1964), a benchmark in pastoral scholarship, clarifies Stegner's ambitions with the pastoral theme, for Stegner's major works are fundamentally concerned with the juxtaposition of machine and garden in the American landscape. These terms are best read with wide meaning, of course: the machine in Stegner's work signifies the various technologies of cultivating land, including irrigating, fertilizing, and planting, and also comes to mean all manner of building and general land development; the garden stands for reclamation or reformation of land in the sweeping sense of Jefferson's agrarianism. The garden is thus a blanket term for fields, orchards, city parks, and even suburban yards, so long as they reflect ideals of land use.

Within this framework, Stegner writes a modern, critical pastoral in the sense that he entertains for a time the prospect of the machine actually gracing the garden. His characters' hopes are all in that direction as they set out to improve nature (and thus their lives) with technology. However, the ultimate thrust of Stegner's writing is that this meliorist pastoral dream is often flawed. Even though alluring, the garden cannot keep its promises. What emerges, then, is an understanding of land that is decidedly ecological. Stegner brings his characters to see that land, especially dry, western land, cannot tolerate the idealized pastoral design. The implicit suggestion, of course, is that we too should begin to respect land.

Stegner's conception of the pastoral appears to depend a great deal on an experience related in his autobiography of 1962, *Wolf Willow*. This work, in fact, serves as a capsule statement of the complex of pastoralist ideas appearing in the novels. In *Wolf Willow*, he shows himself trying—at the very limits of his strength of character and intellect—to hold both nature and technology in some sort of complementary balance. He does this most clearly when, after nearly half a century of living away from his boyhood home, he returns to Whitemud, Saskatchewan, where he walks in search of traces of once having worked with his father to make the prairie garden-like. In a sense, he embarked on a pastoral quest, looking for his identity in the evidence of improved land.

Yet unexpectedly the land had gone back to prairie, and as Stegner wanders over what had been the family farm, he finds reason to give up the search for the footings of barns and the outlines of the fence rows. In their places, he finds that the delicate smell of wolf willow, which had never before recommended itself to his notice, transports him. For the moment he feels himself removed to a realm where "reality is made equivalent with memory, and a hunger is satisfied" (19). With that subtle reassurance of a connection to the land, he no longer feels compelled to dig up the farm. He could let it lie buried, content in knowing that he would not recover his sense of being home in scars he had left on the land.

The significance of this experience is that it helped Stegner decide how he would thereafter treat land. Ultimately, if he had to leave marks, they ought to be no deeper than the foot paths he made as a boy. It came to him that the feel of "wearing" paths into "the earth's rind" was somehow "an intimate act, an act like love . . . denied to the dweller in cities" (275). The marvel of walking out again on the prairie of his youth was that his feet had "printed an earth that seemed creation-new with the marks of [his] identity," and then the land, like a lover, had covered his passing. It is this impulse to use land gently, insuring that it can straightaway return to its natural state, that Stegner develops in his fiction.

The first of the novels germane to the present investigation is *A Shooting Star* (1961), which must be approached as an early and uncertain demonstration of the pastoral/ecological thesis. The problem is that the action is too scattered for the conflict between technology and nature to develop coherently. In addition, entire chapters stray away to a pastoralism that may be more accurately described by Marx's terms "popular and sentimental," rather than "complex and critical." In the mid-section of the novel, for example, there are settings in Mexico and in the Sierra Nevada near Lake Tahoe where the harried protagonist does not attempt to improve land but simply to run from her cares.

Where the action in *A Shooting Star* pertains to land development, two symbols are dominant, the bulldozer and the espaliered fruit tree. These stand for pastoralists' heavy-handed attempts to improve nature. In the bulldozer we see that even though land can be pushed around at will, it should not be gouged deeply when grading roads and foundations. And in the espaliered tree we realize that art is sometimes a cruel form to force on life. For however incongruous bulldozers and espaliered trees might at first seem, their pairing effectively comes to suggest that no matter how beautifully we might landscape, our work is usually destructive of nature. This is the lesson Sabrina Castro learns when she finally acknowledges that "no combination" of garden or park or woodland "really turns out to be the kingdom of heaven" (401). This insight enables her to return to her family origins after having exhausted herself physically and spiritually in the quest for a safe garden hideaway. In a closing scene, we see her steeling herself (somewhat too melodramatically) to her brother's bulldozers that

> will start smashing down . . . and uprooting the oaks, and I'll have to stay around just to see that he maintains some minimum responsibility, and doesn't split it up into little hundred-foot lots for the fastest and biggest profit. (400)

One difficulty here is that Stegner is too willing to slight the story at hand in favor of making an appeal to the political allegiances of his readers. The overly ironic development of the character of Oliver Hutchens, Sabrina's brother, is the principal case in point; one of the first things we learn about him is that he wants "to take advantage of the [real estate] boom" by subdividing the last open tract in the neighborhood (179). Such a desire makes him a "very American character" whose excesses of energy and singlemindedness will "trample" those standing in his way (330).

But rather than let this signal to his reader go at that, Stegner also makes the narrator bombastic. If readers are willing to go along with the heightened rhetoric, they will not only dismiss Oliver, but also will likely decide against anyone else who happens, conscientiously or not, to work in the real estate profession. The effect is a loss of focus, as Stegner goes on to remark:

> His kind never anticipated consequences. His was the kind that left eroded gulches and cutover timberlands and man-made deserts and jerry-built tracts that would turn into slums in less than a generation. They got awards from service clubs and resolutions of commendation from chambers of commerce. They denuded and uglified the earth in the name of progress, and when they lay on their deathbeds—or dropped from the massive coronary that the pace of their lives prepared for them—they

were buried full of honors and rolling in wealth, and it never occurred to the people who honored them, any more than it occurred to themselves, that they nearly always left the earth poorer and drearier for their having lived in it. (321)

The writing here barely conforms to the genre of the novel. In the way that it generalizes, the passage seems instead a polemic, though admittedly the distinctions among genres do not stand much scrutiny. Still, Stegner seems clearly less intent on advancing his plot and more concerned with exposing what can simply be called a bad class of people.

Stegner digresses, it seems to me, for a couple of reasons. The first is that by the time he wrote this novel, he had already produced several essays in the service of conservation.[1] It is reasonable to believe that the rhetoric and perhaps some of the cadence come, at least indirectly, from such pieces as "The Rediscovering of America: 1946," later collected as the second chapter in *The Sound of Mountain Water* (1969), in which Stegner began to write passages like:

> . . . the automobile is the natural enemy of dozens of small creatures— rabbits, gophers, snakes, mice, lizards, ground squirrels. The highways throughout the West are practically paved with flattened, crisped, sun-dried rabbits and ribboned snakes and wafered squirrels. By killing off the coyotes and wolves we let the rabbits and rodents multiply; by building highways we lure them to their death under our furious wheels. "Compensation," says Mr. Emerson. "It all evens up," echoes Mr. Hemingway. I wish I felt it possible to draw a moral about how it does not pay to be a predator in this brave new world, but all I can devise is that it doesn't pay to be anything except man. Maybe it doesn't pay to be man either. (48)

Moreover, we might imagine how difficult it must have been for Stegner to stay within the conventional bounds of the novel when on every side, in the post–World War II boom, people were ruining land. Indeed, the wonder may be that Stegner did not lapse more often from his novel to take jabs at the worst offenders.

Essentially, this is the question that Richard Etulain put to Stegner in the booklength interview of 1983, *Conversations with Wallace Stegner*. Etulain wanted to know to what degree Stegner's characters are spokesmen for issues, especially conservation (171). To this, Stegner answered that he hasn't "put any outright conservationists into my fiction," except for the "little leakage" of "anti-development" sentiment in *A Shooting Star*" (172). Stegner further remarked that, yes, while conservation has been a cause of his, he understands that "it means joining, it means activism, and I don't think fiction should really have proselytizing as its purpose" (172).

Yet, as Etulain rightly sees, Stegner's characters do speak, and often quite explicitly, in the interest of conservation. The resolution to this apparent contradiction is that Stegner finds in the pastoral convention the means to satisfy the demands he placed on himself to produce art. At the same time his version of pastoral admits an interpretation of theme that readers generally see as ecological. In the case of *A Shooting Star*, it is merely a matter of recognizing that Stegner incompletely worked out the pastoralism that more effectively carries the later novels.

Primarily, his difficulty is not setting, but characterization. Sabrina is too passive and confused a creation. Clearly, Stegner drew her out of sympathy for the plight of women. But her character merely suffers from sexual stereotyping; hardly ever does she gather herself against inequities. The result of this weakness is that she cannot fully engage the opposition between nature and technology. In fact, throughout much of the novel, this rather intellectual concern remains beside the point. Instead, overwrought emotions drive her character, producing what James Light calls a "puffed-up, repetitive" soap opera "only slightly removed from the afternoon world of Proctor and Gamble" (1961, 104). Sharp criticism, surely. But Stegner himself, looking back some twenty-six years after creating the character, admits that Sabrina is "soap-operaish" (Etulain 1983, 90).

The character who is more genuinely a creation of the pastoral has, as it happens, a relatively minor role. Leonard McDonald is the one most capable of wrestling with the nature/technology opposition. That he is at once a brutish looking "ape-man" and yet an accomplished scholar and teacher signifies this duality. Thus he is capable of waking one morning,

> listening for the sounds that traditionally meant morning, and heard none of them. No roosters, no milkhorses, no buses or early-to-work trucks rumbling across the bridges of sleep, no chirp of sparrow or hammer of woodpecker, no curtains moving in light auroral breezes with a slur of fabric on wood or screen. What Greenwood Acres awoke to were the

sounds of the over-crowded future, where what is good is what annoys least. Morning in Greenwood Acres was zoned out. (193)

Leonard, far more than any other character, is able to see that suburbs are not the harmonious mixes of nature and technology that their advertisements claim. They are instead fussy and repetitious embellishments of land that rarely, if ever, provide the freedom and dignity people look for outside the inner city. Stegner's criticism of the suburb and land developers is one of the most memorable parts of the novel, for only insofar as Sabrina comes to share Leonard's hatred of his cramped neighborhood does she stop herself from going, as she often says, "to waste."

Giving his protagonists a clearer awareness of the nature/technology dilemma has made a world of difference in the pastoral novels since *A Shooting Star*. It can be said almost categorically that Stegner has not since gone off on tangents such as his all-out attack on the profession of land development. Nor has he delivered himself of more condescending explanations of the damage that bulldozers can do. And yet he has in no way pulled back from showing the consequences of abusing the earth. He has simply found how to set aside the narrow and literal significance of his images and to concentrate instead on a more powerful artistry.

Perhaps it is Stegner's shift to first-person narrative that makes *All the Little Live Things* (1967) and its sequel *The Spectator Bird* (1976, National Book Award 1977) more compelling novels than *A Shooting Star*.[2] In *All the Little Live Things* Joe Allston's admission of wanting only to retire and tend his garden helps make him a universal and familiar character. So too, and in part because of her relationship with Joe, does Marian Catlin seem a genuine and fresh creation. A young woman with a zest for life, she is Joe's neighbor and antagonist, whose presence goads him into revealing his inner conflicts. It is just at her entrance in the story that we learn Joe is capable of a truly amazing blend of nature and technology. With startling irony he outlines his plan for better gardening through chemistry:

> So here I am contentedly sprinkling cutworm bait along my row of young tomato plants and amusing myself thinking what a quaint idea it is to perfect Eden with poison, and wondering (let us suppose) what Adam and Eve did without rotenone, melathion, lindane, chlordane, sodium ammate, and the other deterrents. (56)

This revelation of Joe's character sets up the drama between him and Marian, the first exchange of which leaves Joe musing:

But good Lord, what this charming idiotic woman is saying! She wants to restore natural balances that have been disturbed ever since some Cro–Magnon accidentally boiled his drinking water." (63)

The subsequent debates between these two keep up this half-serious banter, and the novel is remarkable in remaining at once lively and plausible even as it ranges over such weighty philosophies as Manichaeism, Darwinism, Calvinism, and Transcendentalism. Despite what might at first seem a grab bag of ideas, the themes steadily come together in an argument over whether life is mechanistic and stupid, as Joe says, "a simple interchange of protein" (91), or if it's as Marian hopes, an evolutionary "perfecting of consciousness" (168). One of the points of the ongoing argument, expressed by Marian, is that "plants whose genes were adapted to an environment ought to be let live in it" (96). The broader environmental principle is that native plants are superior to exotics, particularly in the desert West, which Joe as a New Yorker is slow to realize.

Joe eventually acknowledges a partial "conversion" on this point, and reflects that he has,

like other immigrants . . . brought the familiar to an unfamiliar place, our planting impulse no different from that of pioneer women hoarding in their baggage seeds of lobelia and bittersweet, or Jonny Appleseed scattering civilization along a thousand miles of frontier tracks. Call it the Law of Dispersion and Uniformity. Marian, who valued the indigenous over the exotic, was almost the only person I ever knew who didn't submit to it, and even she would sometimes take pleasure in the results. (125)

This realization finally helps Allston stop himself from killing "all the little live things" that spoil the "botanical garden" he has tried to force on his yard.

An absorbing study in itself, and one that also lends itself to an ecological reading, is Joe's role reversal from *All the Little Live Things* to *The Spectator Bird* (1976). In this novel Joe does not have to be swayed by an antagonist but speaks immediately and forcefully in the defense of the land. Having been, in his own words, a "curmudgeon," Joe comes back for a second performance with his sense of ecology greatly heightened. During his travels to Denmark, Joe happens on Eigil Rodding, whose experiments in genetics threaten natural order. The "Doctor Faustus of genetics," as his neighbors call him, Eigil has bred his crops and stock to the point that his estate is a model of productivity. Yet Joe sees that Eigil no longer has "a farm, it's an economy," where nature has been "improved" so much that "no contented cows" live there, only "Stakhanovized cows," which if they don't "keep up [their] statistics" end up as "schnitzel" (145–147).

It's here midway through the novel that Stegner fully unites his several themes: the danger of genetics creating monsters, technology displacing people from traditional agriculture, and monocultures killing off the deer, rabbits, birds, and other wildlife that typically make their homes within and beside agriculture. Although Joe finds part of himself attracted to Eigil's fields, he cannot help but notice that

> at the hour we went around, there was hardly a working soul in sight. He had everything mechanized, even automated. The peasants who used to work on the place must all be up in Copenhagen on welfare. . . . (146)

Not many pages after this Joe puts a sharp edge on the truth that he speaks, warning Eigil that his "blueprint" for genetic purity does not allow his crops and stock to keep their "hybrid vigor" (146, 150). That Joe makes it clear he speaks "as an American" suggests that we should take care in this country not to develop our lands as thoroughly as has western Europe.

The conflict ultimately comes down to Joe pointedly telling Eigil that "wild things" are missing "on this marvelous estate of yours . . . little cotton tails or gophers or snakes or moles or raccoons or polecats that could breed in the hedges and live in spite of you" (152). Thus Stegner uses the pastoral conflict to arrive at a conclusion: the health of land can most easily and perhaps most accurately be measured by simply noting how far it has been removed from wilderness. The implication is that certain ideals of wilderness are everywhere desirable, whether or not land happens to have the status of a legislated wilderness.

The notion that land needs protection would have simply dumbfounded the nineteenth-century characters in *Angle of Repose* (1971, Pulitzer Prize 1972). Chief among them, Oliver Ward comes from the East armed with the technology of irrigation. He is confident above all that he can turn the desert into the Garden West.[3] The showpiece of his success will be his "demonstration ranch" near Boise, Idaho. This enterprise, however, presents a stark contrast to the other garden settings of the novel that attract his wife, Susan. Had she not married, almost certainly she would have remained on her parents' small but comfortable farm in Milton, New York. There she would have spent her life perfecting her art, which in its conformity to the "American Cottage Tradition" represents a familiar pastoral form. Or had she a second choice, she would have remained in the hacienda at Michoac'an, Mexico, where the pastoral in the Latin tradition offers the gentility that she dreams of.

But as the wife of a mining engineer, she travels to the roughest towns and camps, where her fate is to be an unwilling partner in her husband's schemes to bring civilization to the West. One of the most potent ironies of the novel, therefore, is that even though both Susan and Oliver are irresistibly drawn to pastoral lives, they subscribe to competing versions of the myth. She faces eastward or "backward" to established gardens, while he faces westward to gardens that he might make, if he works hard enough, from the desert.

For a time Susan sees this conviction of Oliver's in terms of an "absorbing vision":

> He follows it like a man panting after a mirage, and he works, works, works. He manages his survey, he supervises the ditch construction, he confers with politicians and contractors and shareholders, he takes visiting representatives of the Syndicate over the works . . . and in the hours between dusk and dark, he is out with John doing something to the land or the buildings or the well. (424)

Unfortunately for their marriage, however, Susan can share this vision of Oliver's only while living in the relative comfort and prosperity of "The Canyon," where she helps build a home of her own. Then, briefly, she sees the desert as beautiful and understands her husband's enthusiasm for the land (338). Even though she has always hated the West, her letters home to New York begin tentatively to mention the blessings of "Eden" (341) and "Brook Farm" (348), and she, like Oliver, sees good omens in double rainbows (373–374). In a burst of hope, Susan writes to her friend Augusta Drake, who enjoys a refined life in a mansion on Staten Island:

> Have you ever built a house with your own hands, out of the materials that nature left lying around? Everyone should have that experience once. It is the most satisfying experience I know. (348)

She gets so caught up in her home that her sense of time becomes "cyclic, not chronological" (352), suggesting that she too has come to live in the dreamtime of the pastoral, held like her husband by "the freedom, the active outdoor life, [and] the excitement of something mighty to be built" (342).

But most of the time Susan remains a skeptic who deflates Oliver's false hopes. Her love for him does not color her view. She sees few changes to the desert and comes to think of her husband's vision as an "incurable Western disease" that drives him to "triangulate his way across a bone-dry future, dragging her and the children with him, until they all died of thirst" (328).

It's from studying her papers for a family history that her grandson, Lyman Ward, reaches similar conclusions about the possibility of ever making the Garden West a reality. The difference, however, is that nearly one hundred years after the fact, Lyman is able to reach a more objective, kinder assessment of his grandfather's quixotic projects:

> As a practitioner of hindsight I know that Grandfather was trying to do, by personal initiative and with the financial resources of a small and struggling corporation, what only the immense power of the federal government ultimately proved able to do. . . . Like many another Western pioneer, he had heard the clock of history strike, and counted the strokes wrong. Hope was always out ahead of fact, possibility obscured the outlines of reality. (341)

Yet even in Lyman's careful handling of his grandfather's part in early reclamation projects, there is the ecological awareness that his grandmother dimly intuited when she saw that Oliver was not likely to remake the West wholesale. We come across this awareness fairly early when Lyman observes that his grandfather's formula for an improved cement had both "built" and "ruined" the West (168). Stegner makes full use of this observation in designing an irony in the novel: Oliver lives to see his invention of cement help raise the "Arrow Rock Dam" that drowns the one place, "The Canyon," where he had lived happily with his wife (341).

Stegner's most recent use of the pastoral in the defense of land is *Crossing to Safety* (1987). Unique among his novels, *Crossing to Safety* does not concern itself with the use of western landscapes. Rather the story takes its protagonist, Larry Morgan, from New Mexico to the Midwest, and finally to the East. The reason for this shift may be no more significant than that Stegner has served notice to his readers that he chafes under the label of a regionalist,[4] and with this latest release he breaks down what he has called the western "prison" and the "regional peekhole" (Etulain 1983, 124–125).

However, another and more important reason presents itself. Stegner's heightened sense of ecology undoubtedly tells him that not only is the pastoral ideal unworkable in the desert West, but it is also suspect in the East. Because of our much greater ability to interfere with nature and, in effect, to tame land to our purposes, we find ourselves abusing the lusher and more resilient lands eastward just as surely as we wreck the fragile deserts. So we may think of the pastoral theme in *Crossing to Safety* as Stegner's attempt to raise his readers' awareness not just within a region, but across the nation, if not indeed also throughout the globe. Thus we find Larry disturbed that his friend, Charity Lang, uses the woods of New England "in defiance of the genius of this country, which tries always to hide itself in trees" (228). He is particularly perplexed that she has "bulldozed" a meadow to make a playing field for her kids (265). As Larry walks up to what had been "an empty field," he reflects how he has at his feet

an absolutely characteristic Charity artifact. She had prepared that field in a burst of enthusiasm without considering that it was a steep mile and a half from the lake where all the children lived. . . . Obviously nobody had played anything there all summer. (265)

Larry's point can be extended in all directions of the compass, as everywhere people like Charity look on raw nature as "disorder and scarred earth," which each spring season they must again "set to work to transform into a landscape" (179).

This sought-after landscape, as it has been throughout the previous novels, is the pastoral garden. But as Eden was not "valid without [its] serpent" (139), neither is the modern garden a clear improvement of nature. The price of the garden is constant and often grueling labor to preserve the artifice which, for all its beauty, somehow does not satisfy. The best we might do therefore is as Stegner bids in this novel—stop and ponder "the miserable failure of the law of nature to conform to the dream of man" (267).

This treatment of the pastoral/ecological complex cannot be completed without making a distinction between conservation and preservation. A few years ago, Stegner began to break with the ranks of the conservationists. His view has evolved beyond merely conserving land so that future generations can put it to use; he now wants to preserve land, on its own terms, for all time. It is worth pointing out this shift, because such prominent shapers of ecological opinion as Edward Abbey persisted in talking about Stegner as if he were an unreconstructable conservationist. Abbey went so far as to suggest that Stegner forfeited a place among the first rank of those framing our attitudes toward land because of his "excess of moderation, an extremity of forbearance" (1969, 10).

But it seems to me that Abbey looked too steadfastly on Stegner's early reputation as the sort of conservationist who was willing to compromise. Abbey was not very often inclined to compromise, while Stegner has been one who would bend a little. This has made him one of the few champions of public lands who has been able to sit down over drinks with Sagebrush Rebels without coming to blows. There is value in that, and what Abbey dismissed too casually is that Stegner has consistently held the position he took as early as 1969—that wilderness is "the geography of hope" ("Coda: Wilderness Letter" (153). This hope is notably absent in *A Shooting Star*, even though Stegner wrote it at about the same time he turned out the essay. Leonard McDonald's attitude toward land is essentially that of a conservationist. Although Leonard wants to turn back the bulldozers, he has his own designs on the land—a park. Concerned hardly at all about the natural quality of the land, his plans call for well-tended lawns and flower beds. But in light of recent environmental attitudes, Leonard's park would be hardly more acceptable than a parking lot.

By contrast, Joe Allston, Lyman Ward, and Larry Morgan, characters created after Stegner revised his view of land, all share some fundamental apprehension of unchecked development. Nowhere do any of them utter the word "wilderness," but all make quick distinctions between land that remains in its natural state and land that has been put to use. And without exception, all make themselves comfortable with the prospect of land left alone so that nature might go its own way.

Notes

1. See the collected essays in Stegner's *The Sound of Mountain Water*, from Garden City, New York: Doubleday, 1969. Stegner also addresses the topic of ecology in his interview with Richard Etulain, *Conversations with Wallace Stegner on Western History and Literature*, from Salt Lake City, Utah: University of Utah Press, 1983. See the last two chapters, "The Wilderness West" and "What's Left of the West" (167–198).

2. For a discussion of Stegner's use of first-person narrative, see Audrey Peterson's "Narrative Voice in Wallace Stegner's *Angle of Repose*" in *Western American Literature* 10 (August 1975): 125–133.

3. For a discussion of the "Myth of the Garden West," see Henry Nash Smith's *Virgin Land: The American West as Symbol and Myth*, New York: Vintage, 1957. Although Smith's myth/symbol school is no longer popular among some scholars, it nonetheless offers a powerful explanation for the motives of the characters in *Angle of Repose*.

4. For a treatment of the problems that western writers face, see Stegner's "Born a Square" in *Atlantic* 213 (January 1964): 46–50; rpt. in *The Sound of Mountain Water*, 170–185.

References

Abbey, E. Review of "The sound of mountain water." New York Times Book Review. 1969 June 8:10.

Empson, W. Some versions of pastoral. (Rpt. 1968.) New York: New Directions; 1935.

Ettin, A. V. Literature and the pastoral. New Haven, CT: Yale University Press; 1984.

Etulain, R. Conversations with Wallace Stegner on western history and literature. Salt Lake City: University of Utah Press; 1983.

Finney, G. Garden paradigms in nineteenth-century fiction. Comparative Literature 36(Winter):20–33; 1984.

Greg, W. W. Pastoral poetry and pastoral drama. London: Bullen; 1906.

Light, J. F. A motley sextet. Minnesota Review 2:103; 1961.

Marx, L. The machine in the garden: Technology and the pastoral ideal in America. London: Oxford University Press; 1964.

Poggioli, R. The oaten flute: Essays on pastoral poetry and the pastoral ideal. Cambridge, MA: Harvard University Press; 1975.

Stegner, W. E. A shooting star. New York: Viking; 1961.

———. Wolf willow: A history, a story, and a memory of the last plains frontier. New York: Viking; 1962.

———. All the little live things. Lincoln: University of Nebraska Press; 1967.

———. Coda: Wilderness letter. In: The sound of mountain water. Garden City, NY: Doubleday; 1969:145–153.

———. Angle of repose. New York: Fawcett; 1971.

———. The spectator bird. Lincoln: University of Nebraska Press; 1976.

———. Crossing to safety. New York: Random House; 1987.

Steinbeck, J. The white quail. In: The long valley. New York: Viking; 1938:27–44.

A *Wilderness of Rivers*

River Writing in North America

CHRIS BULLOCK

GEORGE NEWTON

Three Wilderness Themes

T H E R E are obviously a variety of wilderness values, reasons why people have gone and continue to go to wilderness. Three values, or themes, seem particularly important. First, and perhaps most complex, wilderness allows one to return to a more primordial and natural state. Laurens van der Post in "Wilderness: A Way of Truth" speaks of "wilderness man," the primitive inhabitant of wilderness, who also "exists in us. He is the foundation in spirit or psyche on which we build, and we are not complete until we have recovered him" (1984, 236). Wilderness is also associated with our own earlier selves, the realm of childhood. The philosopher Alan Drengson, for example, speaks of

> the small wilderness of natural self that is the unique quality of each person. Children are not wild beasts to be tamed, their spirit imprisoned by control, but as natural beings come with their own needs and capacities for spontaneous, creative action. (1986b, 4)

Wilderness's association with the primitive also implies an association with a particular range of qualities. In Zamiatin's *We*, as Wayland Drew points out, wilderness is the realm of irrationality, viewed as a necessary refuge from the oppressive rationality of the 'civilized' world (1986).

For John Fowles, in "The Blinded Eye," freedom depends totally on the freedom of the natural world (1984). Wilderness, then, can be seen as returning us to an earlier, more natural world, where we experience freedom from life-denying external and internal controls.

Another way of valuing wilderness is to see it not as reflecting something in ourselves but constituting a world that is wholly other. Jay Vest argues that "wilderness . . . means 'self-willed land' or 'self-willed-place' with an emphasis on its own intrinsic volition. . . . [the] 'will-of-the-land' conception—wilderness—demonstrates a recognition of land in and for itself" (1986, 4). Tom Birch argues that our relation to wilderness "requires an active respectful acknowledgment of the self-integrity and freedom of the Others with whom, and with which, we are relating" (1986, 23). Dave Foreman in "Dreaming Big Wilderness" argues that wilderness is required not for escape, or for recreation, or for the protection of watersheds, but simply "because it is. Wilderness for its own sake. Because it's right. Because it's the real world, the repository of three and a half billion years of organic evolution" (1986, 26). As Wendell Berry puts it,

> where is our comfort but in the free, uninvolved, finally mysterious beauty and grace of this world that we did not make, that has no price, that is not our work? Where is our sanity but there? (1988, 21)

Because wilderness is other, it can be seen as "sacred space." Precisely because we do not control it, wilderness can be experienced as a "place pervaded by a sense of power, mystery and awesomeness" (Miles 1986, 15).

The third value in wilderness is the awareness it gives us of our connection with the community of beings on Earth. Alan Drengson sees wilderness travel as "a journey back to reality, back to the real world of embodied life within a larger biological community of beings" (1986a, 2). Thomas J. Lyon, commenting on the wilderness pioneer John Muir, concludes that

> John Muir's life and thought showed that the wilderness experience, that strangely potent state we are so hungry for, is simply to relax into and recognize the flow and totality of the wild universe. . . . To speak now in a metaphor of our own time, the wilderness experience is not a trip but a return. We seem to have been away. (1986, 22)

Bill Devall, well-known proponent of "deep ecology," claims that "when we defend wilderness we defend something in our self—our larger Self rather than the narrow self usually considered as our Social Identity. We are defending our connection with the greatness beyond our narrow self" (1986, 24). The awareness of connection that comes from the wilderness experience can change our whole attitude toward land in general. In the well-known words of Aldo Leopold, "We abuse land because we regard it as a commodity belonging to us. When we see land as a community to which we belong, we may begin to use it with love and respect" (1970, xviii–xix).

River Themes

An examination of river writing, to draw out characteristic river values/themes, reveals that these themes have interesting similarities to, and differences from, the general wilderness themes we have just described.

The idea that rivers allow a return to an earlier, freer, and more natural state exists throughout river writing. Nathaniel Hawthorne, for example, describes a fishing excursion on the Assabeth River:

> Strange and happy times were those when we cast aside all irksome forms and straightlaced habitudes and delivered ourselves up to the free air to live like Indians or less conventional beings during one bright semi-circle of the sun. (1971, 14)

For Hawthorne,

> the chief profit of those wild days . . . lay . . . in the freedom which we thereby won from all customs and conventionalism and fettering influences of man on man. We were so free today that it was impossible to be slaves again tomorrow. (1971, 15)

The river becomes a central symbol of freedom in the best novel by Hawthorne's successor, Mark Twain. In *The Adventures of Huckleberry Finn,* Huck travels down the Mississippi in the company of Twain's version of natural man, the Negro, Jim. The river itself is hardly described in the book, but it provides a repeated refuge from an adult society seen as utterly corrupt and virtually worthless. Through the agency of the river, Huck is allowed precisely not to grow up; his adventures are essentially excursions rather than developments (1885).

In the twentieth century, rivers flow through the works of some of the most celebrated writers, from James Joyce's Liffey to the Thames, once "sweet" but now polluted, that runs through T.S. Eliot's Wasteland. Most centrally in the tradition of 'river as a source of return' is Ernest Hemingway's Big Two-Hearted River, in the story of the same name. Hemingway's hero, Nick Adams, shell-shocked by the war, a glorious product of civilization, returns home to his favorite river. In the presence of the river, far from civilization, he stands a good chance of rejuvenation, of a return to an earlier, less corrupted self:

> It was a long time since Nick had looked into a stream and seen trout. They were very satisfactory. . . . Nick's heart tightened as the trout moved. He felt all the old feeling. (1926, 160)

However, if rivers frequently help represent the more general wilderness theme of the return to the primitive and free, rivers very rarely symbolize wilderness as wholly Other from man. Instead, we find, again and again, rivers used to symbolize the nature of human life. Nathaniel Hawthorne, for example, suggests that

> if we remember [the river's] tawny blue and the muddiness of its bed, let it be a symbol that the earthliest human soul has an infinite spiritual capacity and may contain the better world within its depths. (1971, 47)

Around the same time, Henry Thoreau, boating with his brother down the Concord and Merrimack rivers, was insisting that "man's life should be constantly as fresh as this river. It should be the same channel, but a new water every instant. . . . Most men have no inclination, no rapids, no cascades, but marshes, and alligators, and miasma instead" (1849, 132). In 1895, Henry van Dyke was claiming that

> little rivers seem to have the indefinable quality that belongs to certain people in the world,—the power of drawing attention without courting

it, the faculty of exciting interest by their very presence and way of doing things." (19)

Finally, Langston Hughes's poem "The Negro Speaks of Rivers" is based on the refrain "my soul has grown deep like the rivers" (1985, 27). It's a short step from seeing rivers as symbolizing aspects of man's life to the third wilderness theme that we identified: the river as symbol of the larger ecological community. For Chief Seattle, as quoted in Joseph Campbell's *The Power of Myth*, rivers were part of the human community. "The rivers are our brothers. They quench our thirst. They carry our canoes and feed our children. So you must give to the rivers the kindness you would give any brother" (Campbell 1988, 34). For Chief Seattle, of course, all things were equally alive. This is the sense we get from more contemporary river writers. In Norman Maclean's *A River Runs Through It*, the author has a vision of the river that starts from the mundane but moves far beyond it:

> I sat there in the hot afternoon trying to forget the beaver and trying to think of beer. Trying to forget the beaver, I also tried to forget my brother-in-law and Old Rawhide. . . . I sat there and forgot and forgot, until what remained was the river that went by and I who watched. On the river the heat mirages danced with each other and then they danced through each other and then they joined hands and danced around each other. Eventually the watcher joined the river, and there was only one of us. I believe it was the river. (1976, 61)

Naturalist Barry Lopez, standing beside the bend of a river, portrays powerfully the dissolving into the riverine community:

> I have lost, as I have said, some sense of myself. I no longer require as much. And though I am hopeful of recovery, and adjustment as smooth as the way the river lies against the earth at this point, this is no longer the issue with me. I am more interested in this: from above, to a hawk, the bend must appear only natural and I for the moment inseparably a part, like salmon or a flower. I cannot say how much this single perception has dismantled my loneliness. (1979, 26)

By looking at the three wilderness values identified earlier and by relating rivers to these values, we start to see the special features of river values. That is, rivers much more easily symbolize our links to wild nature than they symbolize the otherness of wilderness. Where does this special feature come from? What characteristic of rivers leads to this emphasis of river values? The answer lies in the comparison between the flow of rivers and the flow of life.

This analogy between the flow of rivers and the flow of life is memorably expressed in Thoreau's *A Week on the Concord and Merrimack Rivers* when the young Thoreau, close to the end of the book, records a vision in which

> all things seemed with us to flow; the shore itself, and the distant cliffs were dissolved by the undiluted air. . . . Trees were but rivers of sap and woody fibre. . . . And in the heavens there were rivers of stars, and milky-ways already beginning to gleam and ripple over our heads. There were rivers of rock on the surface of the earth, and rivers of ore in its bowels, and our thoughts flowed and circulated, and this portion of time was but the current hour. (1849, 331)

A somewhat similar version is recorded in Annie Dillard's *An American Childhood*. Here Dillard describes the entry into conscious life through the image of the waterfall:

> What does it feel like to be alive? Living, you stand under a waterfall. You leave the sleeping shore deliberately; you shed your dusty clothes, pick your barefoot way over the high, slippery rocks, hold your breath, choose your footing, and step into the waterfall. The hard water pelts your skull, bangs in bits on your shoulders and arms. . . .
> It is time pounding at you, time. Knowing that you are alive is watching on every side your generation's short time falling away as fast as rivers drop through air, and feeling it hit." (1987, 150)

Equally memorable is Loren Eiseley's description of entering and becoming part of the Platte River:

> For an instant, as I bobbed into the main channel, I had the sensation of sliding down the vast tilted face of the continent. It was then that I felt the cold needles of the alpine springs at my finger tips, and the warmth of the Gulf pulling me southward. Moving with me, leaving its taste upon my mouth and spouting under me in dancing springs of sand, was the

immense body of the continent itself, flowing like the river was flowing, grain by grain, mountain by mountain, down to the sea. (1946, 19)

Everything in wilderness, everything in nature, is as alive and moving as we are. In the nature of things, this lesson about our affinities with wilderness is more easily learned from things which move more continuously and spontaneously than from things which move over such a span of time as to make it less perceptible to us. We delight in the spontaneity of wild animals; we can delight and learn equally from the continuity of flow in rivers.

Wilderness Poetics

If the key lesson rivers have to teach us about wilderness is its aliveness, movement, flow, then the question arises: What is it about good river writing which communicates that experience to us? We can tackle that question by looking at the structure and style in the works of some of these river writers.

While many kinds of overall structure exist in river writing, generally river writing adopts at least a loosely chronological or directly sequential structure. For example, Roderick Haig-Brown organizes a *River Never Sleeps* (1944) by the months of the year, Kenneth Grahame's *The Wind in the Willows*, originally published in 1908, follows the passing of the seasons from summer through winter to spring again, and Andy Russell's *The Life of a River* (1987) traces the history of the Oldman River in southern Alberta from prehistoric times to the present.

The most interesting kind of sequential structure is the river journey. This journey is the basis of such works as Mark Twain's *The Adventures of Huckleberry Finn* (1885), the river section in Hart Crane's *The Bridge* (1933), Henry Thoreau's *A Week on the Concord and Merrimack Rivers* (1849), Hugh MacLennan's *Seven Rivers of Canada* (1961), Ann Zwinger's *Run, River, Run* (1975), and Edward Abbey's *Down the River* (1982). However, a crucial difference exists in these river journey writings. MacLennan and Zwinger follow their rivers systematically, faithfully observing each section as they go. In Thoreau and Abbey, though, descriptions of the river form only part of the work. Following the river in these works creates a series of descriptions and reflections. The reflections start, grow, and then conclude by a return to the river; shortly afterwards a new reflection likewise grows and disappears into the flow of the river. Thus in *A Week on the Concord and Merrimack Rivers*, Thoreau reflects on the priority of wilderness over agriculture, the problem of dogmatic thinking, the status of mythology and religion, the uses of literature, the right attitude toward the state, and the nature of friendship. These reflections contrast with powerful descriptions of the meeting of sky, river bottom, and the river's glassy surface. In *Down the River* (1982), Abbey reflects on the need to question authority, the fallacy of doing good, the question of Thoreau's puritanism, and the validity of Thoreau's preaching on simplicity, interspersing these thoughts with descriptions of the Green River, large breakfasts, and solitary trips up side canyons. Works less directly informed by a river journey, such as Annie Dillard's *Pilgrim at Tinker Creek* (1974), also mix description and reflection.

In the overall structure of Thoreau's and Abbey's writing, the reader assimilates the river experience. This structure conveys a preferred model of thought. In the model, thought does not control nature; rather it partakes of the spontaneity and movement of nature. Essentially, these works are a collection of essays, essays in the Montaignian sense of a particular person exploring his or her own flow of thought and perception. Here, individuality of the person serves as the counterpoint to the particularity of each element of wilderness.

The most interesting stylistic moments in river writing come at the point of putting into language the experiences of movement of flow, of the dissolving of boundaries. Sometimes this transfer of the experience into language does not occur very successfully. Early in his book *Seven Rivers of Canada*, Hugh MacLennan describes the difficulty in defining where one river stops and another begins:

> This brings me to a question which has fascinated me over the past few years, a question which I suppose is really metaphysical. What is a river, anyway? According to the *Encyclopedia Britannica*, a river is any natural stream of fresh water, larger than a brook or a creek, which flows in a well-defined channel. As such, it is a basic geological agent. But though the river carves the channel, its water is always changing—the reason why Heraclitus used a river to illustrate his definition of reality,

"Everything flows." Nobody, as he truly pointed out, can bathe in the same river twice.

This old idea re-occurred to me many times when I contemplated the Columbia Icefield. . . . Incredibly, we must presume that some of the water from the Icefield really goes all the way down to the various oceans. For consider the Nile which floods regularly, and carries its melted snows far north through the hottest and most arid desert in the world, yet still has an abundance of water to discharge into the Mediterranean. Yet it is certain that a vast weight of the Nile water evaporates, and so must a great deal of the water in these northern streams. So once again, what is a river? (1961, 5)

Here MacLennan discusses the difficulties of establishing boundaries, but he does so in a calm, judicial prose that seems to clearly separate observer from observed; MacLennan himself seems to be no part of the cycle of flow, evaporation, and condensation. In the opening to her *Run, River, Run*, Ann Zwinger evidences the same sense of separation.

Beneath the beating of the wind I can hear the river beginning. Snow rounds into water, seeps and trickles, splashes and pours and clatters, burnishing the shattered gray rock, and carols downslope, light and sound interwoven with sunlight. The high saddle upon which we stand, here in the Wind River Mountains, is labelled Knapsack Col on the map, a rim left where two opposing cirques once enlarged toward each other. (1975, 3)

This energetic prose catches the delight in the movement of water, but this movement occurs in a landscape viewed by the separated observer.

In other works, however, the authors convey not only a sense of flow but a decrease in the distance between observer and observed. In the passage quoted from Lopez's *River Notes* (1979), the hawk serves as observer and the human being becomes one object in a fairly static landscape. In Dillard's waterfall passage, the short sharp verbal clauses set everything in motion, both the waterfall and the person entering it. Nevertheless, waterfall and person are separate; the "hard water pelts your skull" (1987, 150).

A sense of separation is less evident in a passage from "In a Jon Boat During a Florida Dawn" by David Bottoms:

If you look around you, as you must, you see the bank dividing itself into lights and darks, black waterbugs stirring around algae beds, watermarks circling grey trunks of cypress and oak, a cypress tree fading under a darker moccasin, silver tips of river grass breaking through lighted water, silver backs of mullet streaking waves of river grass. (1985, 34)

This passage begins with a sense of division "into lights and darks," but the rest of the passage deliberately sets everything in motion and shows how all the elements of the river merge into one another. The "grey trunks" fade into the "silver tips of river grass" which are hardly distinguishable from the "silver backs of mullet."

In the passages quoted from Loren Eiseley's "The Flow of the River" (1946) and Thoreau's *A Week on the Concord and Merrimack Rivers* (1849), we see not only the interrelationship of the things in nature but also the merging of the human observer into the flow. In Eiseley's essay, the single river becomes part of the great river system of the continent. The deliberate movement of his formal sentences convey a sense of the immensity of the flow of which he is part. In Thoreau's passage, the flow belongs not only to rivers but to everything, to the shore, the cliffs, the trees, the rock, and even the stars, and the repetitions in his sentences serve to emphasize the universality of flow.

Conclusion

What Abbey, Thoreau, and Eiseley achieve can be best understood by looking at a distinction that John Fowles makes in "The Blinded Eye." In this essay Fowles argues that "nature is a kind of art sans art; and the right human attitude to it ought to be, unashamedly, poetic rather than scientific" (1984, 84). Essentially he points out that the analytic and classificatory mode of science is only one approach to nature; more necessary, particularly for the average person's commitment to conservation, is the celebration of the identity between man and nature, and this celebration is "much more a matter for art than science" (1984, 84). Fowles's distinction suggests that the structures and styles of Abbey, Thoreau, and Eiseley help establish the identity of man's consciousness and the environment. This distinction also suggests the value of the poetic mode in approaching wilderness.

In this same vein, Thomas J. Lyon asks,

> How do we argue for wilderness, not politically, not socially, but in terms of its essential, inner connection with us? There ought to be a medium of communication, a language, which would convey wilderness in its life quality, without "locking it up" into just another item of politics or consumership. (1986, 20)

Abbey, Thoreau, and Eiseley convey precisely "wilderness in its life quality." Similarly, Wendell Berry argues that an era of "new contact between men and the earth" will only "arrive and remain by the means of a new speech—a speech that will cause the world to live and thrive in men's minds" (1970, 14).

However, Berry follows this comment on the new speech, the speech of poetry, by citing R. H. Blyth's insistence that

> poetry is not the words written in a book, but the mode of activity of the mind of the poet." In other words, it is not only a technique and a medium but a power as well, a power to apprehend the unity, the sacred tie, that holds life together. (*in* Berry 1970, 15)

For a poetic understanding of wilderness, we must do more than understand "the words written in a book" about wilderness. While writers may be a guide to language strategies and to the understandings gained through them, finally, our poetic understanding of wilderness must be individual and must be our own.

A key to this individual understanding appears in a passage from John Muir:

> Most people like to look at mountain rivers. . . . After tracing the Sierra streams from their fountains to the plains, marking where they bloom white in falls, glide in crystal plumes, surge gray and foam-filled in boulder choked gorges, and slip through the woods in long tranquil reaches— after thus learning their language and forms in detail, we may at length hear them chanting altogether in one anthem, and comprehend them all in clear inner vision, covering the range like lace. (1971, 47)

Just as we come to know the universal through a reckoning of the particulars, so, too, do we come to know wilderness through a particular experience of it. Here, Muir, like other river writers, has penned lines about an actual experience in a particular place. While a poetic mediation of wilderness is crucial to helping us understand its values, by itself, of course, this mediation is not enough. Just as a river forms a dialectic between watery current and retaining bank, wilderness comprehension serves as a dialectic between flowing experience and poetic discourse.

References

Abbey, E. Down the river. New York: E. P. Dutton; 1982.

Berry, W. A continuous harmony: Essays cultural & agricultural. San Diego, CA: Harcourt Brace Jovanovich; 1970.

———. The profit in work's pleasure. Harper's (March): 19–24; 1988.

Birch, T. The meaning of wilderness. Trumpeter 3(1):23; 1986.

Bottoms, D. In a Jon Boat during a Florida dawn. In: Conley, C., ed. Gathered waters: An anthology of river poems. Cambridge, ID: Backeddy Books; 1985:34–35.

Campbell, J. The power of myth. New York: Doubleday; 1988.

Crane, H. The complete poems of Hart Crane. New York: Doubleday; 1933.

Devall, B. Wilderness. Trumpeter 3(2):22–24; 1986.

Dillard, A. Pilgrim at Tinker Creek. New York: Harper & Row; 1974.

———. An American childhood. New York: Harper & Row; 1987.

Drengson, A. R. Introduction to this wilderness issue. Trumpeter 3(2):1–3; 1986.

———. Introduction to the last wilderness issue. Trumpeter 3(3):1–5; 1986.

Drew, W. Killing wilderness. Trumpeter 3(1):19–23; 1986.

Eiseley, L. The immense journey. New York: Vintage; 1946.

Foreman, D. Killing wilderness. Trumpeter 3(2):25–26; 1986.

Fowles, J. The blinded eye. In: Mabey, R.; Clifford, S.; King, A., eds. Second nature. London: Jonathan Cape; 1984:77–89.

Grahame, K. The wind in the willows. New York: MacMillan; 1908.

Haig-Brown, R. L. A river never sleeps. Toronto: Collins; 1944.

Hawthorne, N. A fishing excursion. In Woods, R. L., ed. Rivers. New York: World; [1846] 1971:14–15.

Hemingway, E. The Nick Adams stories. New York: Bantam; 1926.

Hughes, L. The Negro speaks of rivers. In: Conley, C., ed. Gathered waters: An anthology of river poems. Cambridge, ID: Backeddy Books; 1985:27.

Leopold, A. A Sand County almanac. New York: Ballantine; 1970.

Lopez, B. H. River notes: The dance of herons. New York: Avon; 1979.

Lyon, T. J. John Muir, the physiology of the brain, and the 'wilderness experience.' Trumpeter 3(2):20–22; 1986.

Maclean, N. A river runs through it, and other stories. Chicago: University of Chicago Press; 1976.

MacLennan, H. Seven rivers of Canada. Toronto: MacMillan; 1961.

Miles, J. C. Wilderness as healing place. Trumpeter 3(1):11–18; 1986.

Muir, J. Mountain rivers. In: Woods, R. L., ed. Rivers. New York: World; [1911] 1971:47.

Russell, A. The life of a river. Toronto: McClelland and Stewart; 1987.

Thoreau, H. D. A week on the Concord and Merrimack Rivers. Princeton: Princeton University Press; 1849.

Twain, M. The adventures of Huckleberry Finn. New York: Penguin; 1885.

van der Post, L. Wilderness: A way of truth. In: Martin, V.; Inglis, M., eds. Wilderness: The way ahead. Middleton, WI: Lorian Press; 1984:231–237.

van Dyke, H. Little rivers: A book of essays in profitable idleness. New York: Charles Scribner: 1895.

Vest, J. H. C. Will-of-the-land: Wilderness among Indo-Europeans. Trumpeter 3(2):4–8; 1986.

Zwinger, A. Run, river, run: A naturalist's journey down one of the great rivers of the West. New York: Harper & Row; 1975.

Ways of Knowing Nature
Scientists, Poets, and Nature Writers
View the Grizzly Bear

TERRELL DIXON

THE literature of the grizzly bear is exceptionally rich and varied. Not only does it cover our earliest Western expeditions to the present, but it also cuts across disciplinary lines to include work by explorers and early students of natural history, by scientists, by poets and fiction writers, and by contemporary nature essayists. This diverse literature of the bear offers us a useful way to understand how we know, to look at key points of contrast and congruency in the ways that scientists, poets, and nature writers go about the work of knowing the natural world.

We can begin with one nature writer's rendition of our traditional separation between ways of knowing. Annie Dillard presents us with an engaging description of the difference between scientists and artists as she discovered them in her childhood. Scientists, she says in *An American Childhood*, were "collectors and sorters." They noticed "the things that engaged the curious mind: the way the world develops and divides, colony and polyp, populations and tissue, ridge and crystal." Artists, on the other hand, "noticed the things that engaged the mind's private and idiosyncratic interior, that area where the life of the senses mingles with the life of the spirit" (1988, 213). Since Annie Dillard's discovery was made in the Cleveland Museum of Art, her immediate reference is to the visual arts. The general distinction, however, applies also to how our culture views the literary artist. The scientist, we say, observes and measures to convey objective physical and statistical data; the poet, the short-story writer, and the novelist imagine to create a personal vision. No such ready-made summary describes nature essayists, but their prominent role both in our understanding of nature and in the literature of the grizzly bear means that we need also to see how they know the world.

Scientific study of the grizzly bear began with the work of William Wright, an early hunter-naturalist who published his book, *The Grizzly Bear*, in 1909. Wright's book comes at the end of what has been called the bear hunter's century, 1820–1920 (Schullery 1986), and it chronicles his own change from hunter to naturalist with rhetorical force:

> In the beginning, I studied the grizzly in order to hunt him. I marked his haunts and his habits, I took notice of his likes and dislikes; I learned his indifferences and his fears; I spied upon the perfection of his senses and the limitations of his instincts simply that I might the better slay him. For many a year, and in many a fastness of the hills, I pitted my shrewdness against his. . . . And then at last my interest in my opponent grew to overshadow my interest in the game. I had studied the grizzly to hunt him. I came to hunt him in order to study him. (11)

Like his contemporary, Enos Mills, Wright sought to rescue the real grizzly bear from "the grizzly of popular imagination" (see Wright, 267). The creature which Meriwether Lewis first described as "a turrible looking animal" was usually seen "as an animal who will attack on sight" and who "roams about seeking for whomsoever he may devour" (in Wright, 234). Wright and Mills wanted not to detract from the bear's formidable ferocity and strength, but to look anew at how the bear really behaves.

Wright differed from Mills, however, in that he worked very hard to become "a really scientific naturalist" (257). He was so successful in this goal that contemporary biologists still consider his book to be "one of the best all-around books ever written on the grizzly bear" (Craighead 1977, v). After he renounced hunting, Wright's research progressed by those multiple observations, repeated trials, and cross-checks which characterize the method of the field biologist. He based his views "on many observations, no one of which was conclusive, but all of which, taken together, were not to be ignored" (207). In his later years, Wright buttressed his research by using his camera as a recording and measuring tool. His descriptions of the complex string and flashlighting systems designed to gain night grizzly photographs at close quarters (matched by the equally complex ways in which the grizzlies tried to dodge this set-up) emphasize how thoroughly he sought to gain objective knowledge about the grizzly.

Wright's quest for objective data on the grizzly has been continued in recent years carried out by various researchers, including the Craig-head brothers and their team. The Craigheads stated that their goal was "to complete a thorough ecological study of the grizzly" (1982, 11), and their use of radio collars and tracking effectively turns the individual grizzly into a "free-roaming electronic instrument of science" (1982, 14). By systematically immobilizing, measuring, attaching radio collars, and tracking the grizzlies of Yellowstone over a number of years, the Craighead team has expanded the body of scientific information on the grizzly. The book account of their study, *Track of the Grizzly*, by Frank Craighead provides detailed information about the grizzly's social organization, feeding habits, seasonal movements, ranges, breeding and mortality. It summarizes the Craigheads' years of careful observation and measurement and serves as one model of how contemporary wildlife biology goes about the business of knowing nature.

In *Bear Attacks: Their Causes and Avoidances* (1985), Stephen Her-rero, another contemporary biologist, analyzed 279 records of aggres-sive meetings between bears and people over a seventeen-year period, seeking "to make bear country safer for both bears and people." Re-counting various confrontations, he categorizes them into "sudden" en-counters and "provoked" encounters. After outlining what a bear might do in each situation, he suggests the best options for a human in each type of encounter.

To move from the field biologists' reports on the grizzly to the liter-ary renditions of the bear at times seems like a step back to that period before Mills and Wright had begun to write. One grizzly bear scene in John Hawkes' recent novel, *Adventures in the Alaskan Skin Trade* (1985) can serve as a good example of this. Uncle Jake, the father of the novel's protagonist, moves from France to Alaska where the first of his many adventures is with a grizzly. He rescues a drunken would-be hunter and kills one of the biggest grizzlies on record:

"Seventeen hundred pounds," said Uncle Jake. "And eight feet tall on his hind legs. With a mind like a man's. And a heart you couldn't stop. And all seventeen hundred pounds pumping with the blood of cruelty and wallowing in the fat of destruction and wrapped and hooded in the fur of savagery." (77)

Uncle Jake designates this bear "His Unholiness" and describes their epic battle in terms which include him smelling on the bear's breath "an actual history of regal savagery, the meat and teeth of his combats, the grass and ferns and baby deer he had devoured and then excreted in all the steaming piles I'd seen" (85). As Hawkes' re-creation of early American tall-tale rhetoric suggests, this bear represents in many ways a literary version of the bear in early exploration narratives and symbolizes Uncle Jake's initiation into life on the Alaskan frontier in 1930.

Other creative literature about the bear goes beyond the frontiersman's preoccupation with the bear's strength to look at more idiosyncratic connections between humans and bears. Some poems, like Leslie Ann Silko's "Story From Bear Country" (1981), develop a theme which occurs frequently in Indian legend: the human who goes off to live with the bears. In the legend, this theme usually unfolds through the story of a young woman abducted and kept as the mate of a bear; she then gives birth to an offspring that is part bear, part man. This myth may well have its origins in those times when the bear and the Indian competed for food, and it almost certainly stems from some key physical similarities between bears and humans. Not only does the track of a bear's hind foot resemble a large human footprint, but the grizzly can also use its forearm and dexterous front claws somewhat like a person and it rises up on its hind legs like a human (Craighead 1982).

In Silko's poem (1981) the mingling of lives and identities stems less from forceful abduction and more from an inward pull—a response to an elusive call. The poem, focused on that moment of transition from human to bear, that time when the "you" of the poem "stopped to look back/and saw only bear tracks/behind you," suggests the seductiveness of life amid the beauty of bear country.

> You will know
> when you walk in
> bear country

By the silence
flowing swiftly between the juniper trees
by the sundown colors of sandrock
all around you.

You may smell damp earth
scratched away from yucca roots
You may hear snorts and growls
slow and massive sounds
from caves
in the cliffs high above you.

It is difficult to explain
how they call you
All but a few who went to them
left behind families
grandparents
and sons
a good life.

The problem is
you will never want to return
Their beauty will overcome your memory
like winter sun
melting ice shadows from snow
And you will remain with them
locked forever inside yourself
your eyes will see you
dark and shaggy and thick.

John Haine's short poem on the same general theme, the mythic movement from manhood to bearhood, demonstrates just how distinctly the
poet's private vision shapes his or her knowledge of the bear. As he
recounts in "The Turning" (1974),

I

A bear loped before me
on a narrow, wooded road:
with a sound like a sudden
shifting of ashes, he turned
and plunged into his own blackness.

I keep a fire and tell a story:
I was born one winter
in a cave at the foot of a tree.
The wind thawing in a northern
forest opened a leafy road.
As I walked there, I heard
the tall sun burning its dead:
I turned and saw behind me
a charred companion,
my shed life.

Puzzling at first, "The Turning" becomes clearer once we see that the turning here refers not only to turning to face another direction, but also to the act of becoming altered, of changing into another state. The two sections render the same incident from different perspectives. The first section is told from the point-of-view of the human watching a bear and the second from the view of the bear, now a bear with very human qualities, looking backward. The crucial three lines at the poem's center emphasize this interchange, stressing that the human who keeps a fire and tells a story merges with the bear "born one winter/in a cave at the foot of a tree." In the last lines with the bear looking back at "my shed life," "a charred companion," the poem presents a harsher version of the moment of transition which is also at the heart of the Silko poem.

In one of Maxine Kumin's two poems about the grizzly bear, the poet's imagination knows the grizzly in a very different way. "In the Park" (1989) is essentially a religious meditation; two stanzas which compare Buddhist beliefs on death and rebirth with the speaker's own Old Testament upbringing are interwoven with the story of Roscoe Black, a man "who lived to tell/about his skirmish with a grizzly bear/in Glacier Park." Her poem concludes with a vision of the bear as embodying a punishing god who frightens "atheist and zealot alike. In the pitch-dark/each of us waits for him in Glacier Park."

Frederick Manfred's novel about the frontiersman Hugh Glass, *Lord Grizzly* (1983), presents a variation of this identification of the grizzly with spiritual concerns. In the novel, Glass is attacked by a grizzly which he finally kills, but he is so badly injured in that battle that his companions abandon him, sure that he will not recover. Glass, however, manages to rescue himself by a series of actions which show his sheer resolve to live, his wilderness resourcefulness, and his kinship with the grizzly. He eats the meat of the dead bear, wraps himself in its skin, and crawls on all fours, despite a broken leg, to a faraway fort. As he crawls, Glass motivates himself by thinking of the revenge he will take when he catches those who have deserted him. He also notices, or perhaps dreams, of a giant, ghostly silvertip grizzly who walks behind him on part of this journey. Manfred foreshadows Glass's struggles by presenting a campfire conversation about the Indian's spiritual regard for the grizzly, and he has the novel suggest that, although the rough frontiersman Hugh cannot articulate it, he too achieves a kind of spiritual kinship with the animal whose skin covers him. As the title, *Lord Grizzly*, and his final decision not to seek revenge indicate, it is this kinship which finally comes to be the meaning of his heroic journey.

Revenge of a different kind activates Halverson, the central character in William Kittredge's powerful short story "We Are Not in This Together" (1984). When a young friend of his is killed by a grizzly in Glacier National Park, Halverson quits his job hauling timber and makes elaborate preparations for a grizzly hunt. Halverson clearly wants to avenge the girl's death, but other elements of the story tell us that his quest will be more complex than that. For one thing, park rangers in Glacier have already killed a bear with a belly full of human hair, and the trip to Glacier will also take Halverson into territory which he used to travel with his father and which he has not entered since his father's death. He will also take this hunting trip and spiritual quest with Darby, the woman with whom he lives but with whom he cannot talk.

Once he is in the park's backcountry, Halverson separates his bear hunt from what the rangers have already done; when the rangers fly over their campsite, he feels that "we are not in this together." His own deeper quest is not to be easily realized, however. Halverson kills one bear with his high-powered rifle only to find that death too distant for his purposes. He stalks the second bear with only a knife; he eats the purple-red berries that the bear feeds upon and he comes close enough to smell the bear's "odor of clean rot in the sunlight." When Halverson starts to close with the bear, now standing on its hind legs, Darby shoots it. Just as he and the bear are in it together, eating the same food, stalking each other, and confronting each other, so, finally, is Halverson in it with Darby. By letting her shoot the bear which would kill him, he tacitly acknowledges their partnership. The bear hunting in this complex story thus finally becomes more than revenge; it is also the means by which the half-person, Halverson, can move toward possible wholeness, can reenter the worlds of the park, nature and family, and a life of trust with Darby.

A final element of how creative literature views the grizzly bear can begin with Gary Snyder's "this poem is for bear" (1960). The poem's center suggests a now familiar theme: an Indian girl is abducted by a bear and gives "birth to slick dark children with sharp teeth." The poet-speaker's announcement that "As for me, I am a child of the god of the mountain" identifies him with the offspring of the bear-woman union; once again, bearhood, humanness, and the creativity of the storyteller become bound together. There is, however, another and new element in this poem—one that is especially important to understanding the points of congruency as well as the contrasts in how poets and scientists know nature. One section of Snyder's poem says of the grizzly:

> A bear down under the cliff.
> She is eating huckleberries.
> They are ripe now
> Soon it will snow, and she
> Or maybe he, will crawl into a hole
> And sleep. You can see
> Huckleberries in bearshit if you
> Look, this time of year.
> If I sneak up on the bear
> It will grunt and run.

We may wonder at first how this descriptive catalogue of bear activities—eating ripe huckleberries, hibernating, defecating, grunting, running from a human's approach—fits Snyder's poetic task by serving an ecological, educational purpose. By joining his version of the bear-mating-with-the-young-woman myth to this description of an ordinary bear's ordinary activities, Snyder unites the actual grizzly on the tundra with the grizzly of myth. The importance and worth of the real bear thus become underlined. This emphasis also prepares for the humorous final section of the poem, the one which begins with the provocative "I think

I'll go hunt bears." The humorous, mocking retort to this, "hunt bears?/—why shit Snyder/You couldn't hit a bear in the ass/with a handful of rice," effectively mocks the impulse to kill bears and, with its reference to rice-throwing and thus to marital celebrations, reminds us once again of the human-bear kinship celebrated earlier in the poem.

Snyder's stanza describing the bear's activities also provides a different way for literature to know the bear. His method here—to push his poem toward a less purely private vision of the bear—connects with the methods of other recent poems about the bear. David Wagoner's poem, "Meeting a Bear" (1976), is one such example.

> You may wind up standing face to face with a bear
> Your near future
> Even your distant future, may depend on how he feels
> Looking at you, on what he makes of you
> And your upright posture . . .
> Gaping and staring directly are as risky as running:
> To try for dominance or moral authority
> Is an empty gesture
> And taking to your heels is an invitation to a dance
> Which, from your point of view, will be no circus.
> He won't enjoy your smell
> Or anything else about you, including your ancestors
> Or the shape of your snout. If the feeling's mutual,
> It's still out of balance:
> He doesn't *care* what you think or calculate; your disapproval
> Leaves him cold as the opinions of salmon.

Wagoner's briskness, his humorous comparisons—"a dance which. . . will be no circus," "cold as the opinions of salmon"—enliven, enrich, and encode with his own distinct tone the essentially accurate advice about human-grizzly interaction. If Snyder's description has some affinity with a field naturalist's notes on grizzly behavior, Wagoner's moves toward a fast-paced rendition of what the biologists tell us not to do in a grizzly encounter.

Maxine Kumin's second poem about the bear takes a similar approach. This poem, entitled "You Are in Bear Country," occupies a key place as the poetic prologue to her collection called *The Long Approach* (1986). It is thus an important poem for her as well as for our own efforts to understand some of the complex relationships between bears and humans in contemporary literature.

> They've
> been here
> for thousands of years.
> You're the visitor.
> Avoid encounters. Think ahead
> Keep clear
> of berry patches
> garbage dumps, carcasses.
> On wood walks bring
> noisemakers, bells.
> Clap hands along the rail
> or sing
> but in dense bush
> or by running water
> bear may not hear you clatter.
> Whatever else
> don't whistle. Whistling
> is thought by some to imitate
> the sounds bears make when they mate.

The poem continues with advice on how to act if a bear actually attacks, offering concepts which would not be out of place in Stephen Herrero's book:

> As a last resort you can
> lay dead. Drop
> to the ground face down
> In this case
> wearing your pack may shield your body from
> attack.
> Courage. Lie still. Sometimes
> your bear may veer away.
> If not
> bears have been known

> to inflict only minor injuries
> upon the prone.

In its final stanza, Kumin's poem expands its focus. After exploring how to avoid bear encounters and then how to act if one is unavoidable, she puts our fears in a larger context. By comparing those dangers to the danger of death by bomb, her concern with human-bear interactions broadens into a bleakly existential question, a question which she answers with a combined invitation and ecological statement.

> Is death
> by bear to be preferred
> to death by bomb? Under
> these extenuating circumstances
> your mind may take absurd
> leaps. *The answer's yes.*
> Come on in. Cherish
> your wilderness.

Kumin's poem which contains the subhead "Advice from a Pamphlet published by Canadian Minister of the Environment," shows her, like Wagoner and Snyder, working the borderline between two ways of knowing: the biological and poetic, the scientific and the literary.

This crossing of boundaries between the two traditional ways of knowing has its corollary in the work of the scientists. Just as the poets can know nature through careful observation and through the work of the field biologists, the scientists often see their work as based in the imagination. Stephen Herrero, for example, says about the importance of the grizzly (Schullery 1986, vi): "We should preserve grizzly bear populations, not because their ecological function is critical but because of what they can do for human imagination, thought, and experience." In another example Frank Craighead (1982) dedicates his book to "kindred traits—the inquisitive nature of the grizzly and the imaginative spirit of man."

Craighead and Herrero also share some of the techniques of the creative writer by engaging their readers in popularized accounts of their scientific findings. Herrero's book relates in detail many of the dramatic short narratives about actual human-grizzly encounters, which are the basis for his study, thereby drawing us into the tension of these situations. Craighead's book involves the reader in his scientific study of the bear by discussing several bears whose lives the narrative follows for several years, and especially by focusing throughout the book on a bear that he calls Marian. By thus naming her, by describing significant events in her life, such as mating and motherhood, over a period of years, and by describing her death, Craighead evokes a feeling for her somewhat like that felt for the central character in a novel.

This common ground between literature and science is, of course, often part of the nature writer's chosen territory. It is no accident that Annie Dillard comments so effectively on her discovery of both the scientific and the artistic cultures. Increasingly in our society, it is the nature writer—Annie Dillard, Barry Lopez, John McPhee, or Ann Zwinger—who knows both worlds and who successfully seeks to align and integrate them. Barry Lopez, for example, or Annie Dillard rely on the creative imagination and the techniques of the literary artist, but he or she also shows concern for the kinds of knowledge that the scientist seeks. We can see this blending of the two cultures in essays on the grizzly by John McPhee and by William Kittredge. These essays also show why the nature essay remains so resistant to any easy summaries about how it works.

McPhee's discussion of the grizzly occurs in his book about Alaska, *Coming into the Country* (1977). The first of three sections in the book is called "The Encircled River"; it is a long self-contained essay organized around McPhee's trip down the Salmon and Kobuk rivers. From first paragraph to final pages, the essay expresses his fascination with the bear. McPhee, like many others who have thought and written about the importance of wilderness, acknowledges the bear as the essential symbolic center of our truly wild country.

> The sight of the bear stirred me like nothing else the country could contain. What mattered was not so much the bear himself as what the bear implied. He was the predominant thing in that country, and for him to be in it at all meant that there had to be more country like it in every direction and more of the same kind of country all around that. He implied a world. (62–63)

The first bear sighting, preceded by much apprehension, anticipation and discussion, occasions an essay within the larger essay which deals exclusively with the grizzly. McPhee, who presumably gets some of the scientific information in his essay from the members of his party which includes a Park Service planner, a member of the Bureau of Outdoor Recreation, a wildlife biologist, and a habitat biologist, begins with a description of the bear's size and power. "The immensity of muscle seemed to vibrate slowly—to expand and contract, with the grazing. Not berries alone but whole bushes were going into the bear" (McPhee 1977, 58). This leads to an estimate of its size—"big for a barren-ground grizzly" (58)—and to an explanation of what science has discovered about differences between grizzly size in this part of the Arctic and elsewhere. Next he provides this dialogue:

> "What if he got too close?" I said.
> Fedler said, "We'd be in real trouble."
> "You can't outrun them," Hession said. (McPhee 1977, 59)

McPhee then relates that the grizzly is "no slower than a racing horse" (59) and "half again as fast as the fastest human being" (59) and continues with a skillful summary of other biological aspects of the bear. The narrative follows this mode as it weaves a carefully structured tapestry of first-person experience, Eskimo beliefs, hunting stories, old adages, quotations from one-time, grizzly-hunter-turned-bear-photographer-and-naturalist Andy Russell, and scientific information about the grizzly's eyesight, feeding habits, and denning and hibernation behavior. McPhee also refers to the historical shift among outdoors people from fear of the bear, the kind of fear which prompts the desire to kill it, to reverence, the kind of reverence which prompts the wish to preserve conditions under which the grizzly will flourish.

Throughout this section, other grizzly sightings, fresh grizzly scat, and large grizzly footprints all keep the grizzly's presence uppermost in the reader's mind as well as in the minds of the trip members. McPhee concludes his account with an engaging, complex description of one last grizzly, first as it appears by itself on the Arctic tundra and then as it responds to their approach. At first view, the bear is playing with a ten-pound salmon. "He played sling-the-salmon. With his claws embedded near the tail, he whirled the salmon and then tossed it high, end over end. As it fell, he scooped it up and slung it into the air." McPhee's ongoing description evokes the bear's dexterity, his awesome power, his playfulness, and how these qualities are grounded in his sense of being at home in his land. Then, just as the grizzly becomes bored with this play, their boat comes into view. The bear moves away with "a hurry that was not pronounced but nonetheless seemed inappropriate to his status in the situation." After a brief wait, facing the humans, the bear, "breaking stems to pieces" (1979, 90) disappears into the willows. Here, as so often in his essays, McPhee makes the narrative line speak with symbolic force.

William Kittredge's "Grizzly" is included in his collection of essays entitled *Owning It All* (1987). Where McPhee uses the linear movement of his journey down the rivers to structure his narration, Kittredge's essay is more broadly meditative. It opens with an imaginary dream sequence presenting in narrative form our fears of a grizzly attack in our tents and then weaves together various sources of information—personal, scientific, historical, political—to illuminate "the tangle of emotions" that the grizzly arouses in him and in us. He writes about the bears of Montana, rather than Alaska, and he, like McPhee, feels the importance of this "great and sacred" animal. Wilderness, Kittredge argues, "must indeed, by definition, be inhabited by some power greater than ourselves," and he asks if we can even have "a true North American wilderness without the great and dangerous and emblematic bear out there" (131).

Kittredge's essay, like his short story, seems to originate in the death of his friend who was killed by a grizzly in 1976 in Glacier National Park. Here, however, the friend, Mary Pat Mahoney, is named, and her death leads Kittredge into a discussion of the science, history, and politics of the grizzly in Glacier and Yellowstone. He relies on the views of a more experienced field observer of bears, one who operates outside the official scientific wildlife management community but who has ten years of trailing and filming the grizzly. This friend, Doug Peacock, served as the basis for Edward Abbey's character of George Hayduke in *The Monkey Wrench Gang*, but Kittredge carefully renders him as a different kind of character. As the two talk high atop a lookout tower in Glacier National Park, Peacock talks about being charged by grizzlies maybe forty times, about the bears' "displacement behavior"—something for it to do while it backs down gracefully—about the bear's "critical attack distance"—the fifty yards in which the bear will feel threatened and automatically attack—and about how important it is to give the bear some time and space, to know and use the fact that grizzlies are "always evaluating social and predator-prey signals and relationships" (129).

Kittredge also briefly charts the history of our relationship with the bear. He summarizes Lewis and Clark's response to it with a literary artist's eye for metaphor, emphasizing their description of the Indian preparation for a grizzly hunt "as when they make war on a neighboring nation." He outlines the history of the grizzly's destruction in California by 1922, Utah by 1923, Arizona by 1930, New Mexico by 1931, and one last grizzly—the only one in thirty years—in Colorado in 1979. He weaves the killing of the grizzlies with the pattern of grizzly attacks on humans since 1967, and with discussions of the Craigheads' scientific studies of the grizzlies and the political complexities of bear management decisions.

Kittredge, like McPhee, ranges broadly in his discussion of the bear. It would be a mistake, therefore, to limit what either essay does to simply a blend of scientific information and literary skill. Here, as in other kinds of nature writing, the very inclusiveness of the nature essay precludes any easily agreed-upon generalities about method or about the kinds of information included. Nonetheless, in these two essays, as in much of the writing about the grizzly and other aspects of nature, these two elements constitute a central part of the writing. This confluence of scientific information and literary skill is a crucial element in our increasing interest in nature writing. Its increase in popularity clearly comes in part from our growing environmental awareness (an awareness which is in turn heightened, in a mutually reinforcing way, by the exceptional quality and quantity of recent nature writing). It also stems from an increased specialization in our culture which (despite examples to the contrary here) effectively separates science and literature, driving each to address ever more isolated audiences. Successful nature writing often works against this trend; it lets the reader experience within a single text two central ways of knowing about nature: the objective authority of science and the imaginative power of literature.

In looking at how we know nature, we must not lose sight of the creature which inspires this remarkable range of writing. The literature of the bear in all its diversity and excellence constitutes an eloquent testimony to the grizzly's importance. This importance, however, in no way guarantees that the grizzly will always be there to know. With the grizzly bear, as with the wilderness it represents, a crucial activity—one which must go hand in hand with our measuring and our imagining—is to make sure that the grizzly survives.

Some thinking about the grizzly assumes that ensuring its survival will be a relatively easy task, that the threatened grizzlies of the lower forty-eight states and the currently, much larger populations in Canada and Alaska exist under absolutely different conditions. The population here is vastly diminished, barely surviving in the Bob Marshall Wilderness and in Glacier and Yellowstone national parks. The environmentally minded want to preserve this population, probably because, as Aldo Leopold wrote in 1949, "relegating grizzlies to Alaska is about like relegating happiness to heaven; one may never get there" (1968). We also want to feel that the grizzly in Alaska is still as secure as it seemed in Leopold's time, that the wealth of habitat and of grizzlies in the north is somehow so vast as to be beyond significant harm (Gilbert 1987). This sense of "it can't happen there" should make us uneasy.

We need, instead, to consider that this same attitude once characterized those who took comfort from the bear populations of Yellowstone and Glacier national parks even as they watched the grizzly disappear from elsewhere in the West. For all their individual strength, their apparent abundance and seeming safety, the grizzly bears of the north can, over time, become as threatened as the bears of Montana now are. Pipelines, oil spills, hunting pressures, poaching in the parks, population growth, increased recreational and tourist activity—all suggest that we must pay careful attention to conservation measures for Alaskan as well as Yellowstone grizzlies. Many environmentalists would now argue that the grizzly has its own right to exist independent of whatever scientific, spiritual, aesthetic, and educational value it has for humans. Even for those who stop somewhat short of such a view, however, the great bear's place in our culture, history, literature, and imagination make it clear that we must protect all grizzly populations. Whether we value the bear for itself or for our own observation, inspiration, and wonder, we need to plan and act carefully if we are to insure that the grizzly survives.

[Editor's Note: It should be mentioned that the Craigheads' management recommendations for grizzly bears are highly controversial within the scientific community. Furthermore, a special unit, the Interagency

Grizzly Bear Study Team, has been formed to deal with research and management of this wide-ranging carnivore.]

References

Craighead, F. Foreword to William Wright's The Grizzly Bear. Lincoln: University of Nebraska Press; 1977.

————. The track of the grizzly. San Francisco: Sierra Club Books; [1979] 1982.

Dillard, A. An American childhood. New York: Harper and Row; [1987] 1988.

Gilbert, B. Our nature. Lincoln: University of Nebraska Press; [1986] 1987.

Haines, J. The stone harp. Middletown, CT: Wesleyan University Press; 1974.

Hawkes, J. Adventures in the Alaskan skin trade. New York: Simon and Schuster; 1985.

Herrero, S. Bear attacks. New York: Nick Lyons Books; 1985.

Kittredge, W. We are not in this together and other stories. St. Paul, MN: Graywolf Press; 1984.

————. Owning it all. St. Paul, MN: Graywolf Press; 1987.

Kumin, M. The long approach. New York: Viking/Penguin; 1986.

————. Nurture. New York: Viking/Penguin; 1989.

Leopold, A. A Sand County almanac. Oxford, England: Oxford University Press; [1949] 1968.

Manfred, F. Lord grizzly. Lincoln: University of Nebraska Press; [1954] 1983.

McPhee, J. Coming into the country. New York: Farrar, Straus, and Giroux; [1976] 1977.

Schullery, P. The bears of Yellowstone. Boulder, CO: Roberts Rinehart; 1986.

————. The bear hunter's century. New York: Dodd, Mead, and Company; 1988.

Silko, L. A. Storyteller. New York: Seaver Books; 1981.

Snyder, G. Myths and texts. New York: New Directions; 1960.

Wagoner, D. Collected poems 1956–1976. Bloomington: Indiana University Press; 1976.

Wright, H. W. The grizzly bear. Lincoln: University of Nebraska Press; [1909] 1977.

Why Don't They Write About Nevada?

ANN RONALD

D U R I N G the past century and a half, a distinctly American literary genre called "wilderness writing" has emerged. Henry David Thoreau introduced it; John Muir refined it; hundreds of followers now write variations on the theme. Such authors supposedly are addressing man's relationship to any environment largely untouched by men, or at least that is the common perception. Of course anyone who knows Thoreau's work realizes that Walden Pond was only a couple of miles from Concord, and that Thoreau not only went to town regularly but spent much of his time in the company of other men. His most powerful prose, however, locates a narrator in the midst of an untracked, pristine landscape. Such is the pattern of Muir's work, too. His finest essays extol splendid isolation in the Sierra Nevada, but a quite different reality included the luring of hoards of tourists to Yosemite in order to preserve the valley.

So American nature writing, from its inception, has been characterized by paradox. One might even ask—since the mere presence of a narrator necessarily precludes the existence of true wilderness—whether the genre ever existed in the first place. This "falling tree in a silent forest" puzzle is not my real question, though. My concern here focuses on a slightly different conundrum. Why, as wilderness writing has developed in the twentieth century, are so many authors writing about the same places? And, tangentially, why are they choosing locales that no longer resemble genuine wilderness? Do we prefer only "wilderness" to which human beings can relate? Are contemporary nature writers actually as anthropocentric as those they would condemn?

Glen Canyon serves as a model. John Wesley Powell described it initially as a "curious ensemble of wonderful features—carved walls, royal arches, glens, alcove gulches, mounds, and monuments. From which of these features shall we select a name?" he asks himself. "We decide to call it Glen Canyon. Past these towering monuments, past these mounded billows of orange sandstone, past these oak-set glens, past these fern-decked alcoves, past these mural curves, we glide hour after hour, stopping now and then, as our attention is arrested by some new wonder" (1987, 232–233). Here is the first stage of wilderness writing—the Adamic naming of the place, the uniqueness of the perception, and the fresh language used to picture it. Once Powell framed this initial vision, no subsequent author will have quite the same opportunity to capture an untouched Glen Canyon in words.

Many will try, however. A variety of compelling paragraphs describing aspects of the canyon can be found in Eliot Porter's pictorial version, *The Place No One Knew* (1968), where the photographer presents a visual montage accompanied by excerpts from appropriate wilderness writers. A typical voice is Charles Eggert's: "The face of the cliff was stained with long, black streamers from the water which cascaded over the rim in wet weather. It was an imposing sight, a gigantic backdrop—a motionless hanging tapestry" (Porter 1968, 70). Paging through similar passages reminds the reader that only a finite number of adjectives and nouns appropriately reveal the canyon's magnitude. "On one side above me the red and gold wall was streaked with organ pipes of black and rose and taupe, and on the other, a drift of fringed veil hung delicately purple across its topaz face" (Sumner *in* Porter 1968, 72).

Even as fine a stylist as Edward Abbey, whose *Desert Solitaire* (1968) contains a chapter depicting his leisurely float through Glen Canyon's last days—"Down the river we drift in a kind of waking dream, gliding beneath the great curving cliffs with their tapestries of water stains, the golden alcoves, the hanging gardens, the seeps, the springs where no man will ever drink, the royal arches in high relief and the amphitheatres shaped like seashells" (70)—, cannot perceptibly improve upon Powell's "curve that is variegated by royal arches, mossy alcoves, deep, beautiful glens, and painted grottoes" (1987, 230). Arches, alcoves, glens and grottoes, walls and tapestries, curves and cliffs repeat themselves, until the mid-1960s. Then, suddenly, Glen Canyon changed into a figment of the imagination.

"The place no one knew" became Lake Powell. In telling phrases, a Reno newspaper clipping describes what happened—the dam that was completed in 1963, the subsequent 161,390 surface acres of water, the 1.1 million kilowatts of electricity shared by eight states, the 2,000-mile shoreline, the awesome power of a government that could transform the desert. "Thanks to that Congress, an area was born so different, so colorful, so surrounded by huge colorful monoliths, towering cliffs, spires and peaks that, to movie makers and others, it would take on the appearance of another planet" (Reno Gazette-Journal 1988, 17c). While most wilderness readers will be startled to learn of Congress's omnipotence, we should not be surprised to discover the prosaic kind of wilderness writing such a creation can engender. "Today, with its myriad colors, rugged rock formations and blue skies and colorful reflections, the crystal-clear lake and collection of canyons, buttes and mesas is a vacationer's Shangri-la, one of mother earth's most unusual spots." The lack of adjectival imagination found in this newspaper article is "colorfully" characteristic. I quote from its single-sentence paragraphs only to demonstrate what can happen when a so-called wilderness turns completely civilized. Even Gannett readers are attracted to the place.

On the other hand, not everyone succumbed to the newly-wrought grandeur. Wallace Stegner, for example, boldly negates what has occurred. "In gaining the lovely and the usable," he explains, "we have given up the incomparable" (1985, 128). Twin essays, written seventeen years apart and later published side by side in *The Sound of Mountain Water*, juxtapose the old Glen Canyon and the new Lake Powell. The first recounts a 1947 float trip past "the sheer cliffs of Navajo sandstone, stained in vertical stripes like a roman-striped ribbon and intricately cross-bedded and etched," past "the pockets and alcoves and glens and caves" (117). The second, "Glen Canyon Submersus," acknowledges that "Lake Powell is beautiful." But Stegner's twenty-year-old perception, while reporting that "enough of the canyon feeling is left so that traveling up-lake one watches with a sense of discovery as every bend rotates into view new colors, new forms, new vistas" (126), poignantly concludes that "the protection of cliffs, the secret places, cool water, arches and bridges and caves, and the sunken canyon stillness" (136) have been lost.

The language of this particular wilderness has not been lost, however, for even as canyon country authors were losing Glen Canyon they were already branching into other hidden curves of waterstained red rock. Where John Wesley Powell and Everett Ruess once explored, Edward Abbey followed, trailed closely by Ann Zwinger, David Douglas, Rob Schultheis,[1] and countless others—so many others, in fact, that their terrain begins to look familiar and what once was wilderness has become commonplace. Using the imagery of their predecessors, they simply discover new arches, alcoves, glens and grottoes. One magnificent canyon may be gone but hundreds more remain, each waiting to be described by one intrepid explorer or another, each waiting to be pictured in the same words, more or less.

Perhaps the only wilderness left on the Utah/Arizona border is the wilderness of self. Anyone who has spent a waterless day under the hot sun may disagree, but from a literary point of view, at least, red rock wilderness is no longer a pathless way. In fact, its tracks are so well trodden that the only genuinely new terrain is some inexplicable spot where an inner landscape takes precedence over the outer. Ann Zwinger's solo hike down the Honaker Trail is typical. She begins by acknowledging her uneasiness, "a tinge of apprehension. I'm not quite sure what I expect, and now that my ride is long gone and it's too late and I'm committed, even why I wanted to do this. But here I am, quite alone, and I should start" (1986, 197). Her words could have been written by any one of two or three dozen recent southwest narrators.

Gradually her discomfort eases, her pace slows, and the tone of her prose grows contemplative. Like many of her contemporaries, she ponders the meaning and context of her isolation. "This is what wilderness is to me," she decides, "being alone and knowing no one is within miles, and that although others may have passed here there is minimal, or no, trace of their passage" (210). Apparently she has forgotten what she wrote a few pages earlier, when she described walking past the large yellow letters and numbers left by a Shell Oil Company geological trip. She has not forgotten, however, the subconscious trappings of the civilized world. Again like so many other recent writers, Zwinger balances the wilderness and the city together, noting first "that the materialistic agenda of everyday life does not pertain here," and finally focussing on the self in relation to what she has left behind and what she has found. "Many of us need this wilderness as a place to listen to the quiet, to feel at home with ancient rhythms that are absent in city life, to know the pulse of a river, the riffle of the wind, the rataplan of rain on the slick-rock" (210). Here, then, is the pulse behind so many portrayals of man in the wilderness—Thoreau's "We need the tonic of wildness" (1966, 209–210), for example. The narrator, drawing sustenance from his or her experiences, communicates that physical and spiritual rejuvenation to the reader.

On the next page of her essay, though, Zwinger actually questions the genuineness of the twentieth-century wilderness experience. Does wilderness, in the true sense of the word, any longer exist? Or are the pressures of civilization encroaching irrevocably? Quoting an archae-ologist, Dr. William Lipe, and referring specifically to Grand Gulch, she acknowledges that "we are moving into an era of managed remoteness, of planned romance" (1986, 211). She and Lipe refer to the physical overcrowding of the popular slickrock country, but she might well be talking about wilderness writing, too.

To call contemporary wilderness writing a genre of "planned romance," to consider its practitioners artists of "managed remoteness," levels a serious indictment against the very experience and the atten-dant literature so many readers enjoy. Jokes—about whether this genre exists or not—aside; a serious summary of its extant qualities leads to some telling points. Thanks to the spectacular vistas of certain areas like the Sierras, slickrock country, or Cape Cod, thanks to certain en-vironmental confrontations, and thanks to the potent pens of certain authors, many contemporary wilderness writers retrace each others' steps. Glen Canyon, first described so vividly by John Wesley Powell and later "rediscovered" by a host of successors, is a model environ-ment of a landscape literally overused. That it also has been the focus of man's unfortunate anthropocentrism, developing as it has into a meta-phor for all the dam ills in the West, only heightens the frequency of its invocations. If Glen Canyon indicates a pattern prevalent in the twen-tieth century, then no one really is writing wilderness essays about the wilderness at all. Rather, contemporary authors are writing contempla-tive pieces about "beautiful" places where the hand of man—at least the existence of men—has always been apparent. Or where, if the hand of man is not obvious, the head of the narrator is.

The focus, no matter how splendid or how remote the terrain, centers not on the landscape itself but on the man or woman who visits there. As Everett Ruess's last letter reveals, "I have not tired of the wilderness; rather *I* enjoy its beauty and the vagrant life *I* lead, more keenly all the time. *I* prefer the saddle to the streetcar and star-sprinkled sky to a roof, the obscure and difficult trail, leading into the unknown, to any paved highway, and the deep peace of the wild to the discontent bred by cities. Do you blame *me* then for staying here, where *I* feel that *I* belong and am one with the world around *me*?" (Rusho 1983, 178– 179) [emphases added]. Everett Ruess, wanderer of slickrock country, disappeared near Escalante in November, 1934. Since then, he has become a legend for those who would discard responsibilities, journey alone into a wilderness, and write. His anthems have taken on a kind of mythic quality, drawing others to the landscape he so admired. A source of inspiration, as much a metaphor as Glen Canyon, Ruess here epitomizes the ironic anthropocentricity of a man who most wanted to put away his man-centeredness. The final line of his well-known "Wilderness Song"? "*I* shall sing *my* song above the shriek of desert winds" (Rusho 1983, 181).

It appears that wilderness writers can be as egotistical as anyone else. Meanwhile, their essays repeatedly civilize the very landscape they mean to portray as untracked and unchanged by men. No matter what they write about or why they write, it is clear that the human response is key. And if a wilderness is aesthetically inhospitable to man, American nature writers lose interest. 'Why don't they write about Nevada?' the title of this article asks facetiously. Some images and themes from essays about the Great Basin not only answer this question but point to some important conclusions about the choices wilderness writers seem to make.

Obviously they like Glen Canyon better. As one might surmise, not many Nevada essays exist.[2] Most of the early ones were written by soldiers, adventurers, and pioneers. Just as John Wesley Powell set a standard for the Colorado River country, so the explorer John Charles Fremont crisscrossed and then initially described the basins and ranges east of the Sierra mountains. Fremont's passages, however, are starkly impersonal. "I started out on the plain. As we advanced this was found destitute of any vegetation except sage bushes, and absolutely bare and smooth as if water had been standing upon it" (1956, 447). Even the discovery of a setting as spectacular as Pyramid Lake engendered only moderate effusiveness. "The shore was rocky—a handsome beach, which reminded us of the sea. On some large granite boulders that were scattered about the shore, I remarked a coating of a calcareous substance, in some places a few inches, and in others a foot in thickness" (339). Obviously Fremont lacked the poetic eye of a Powell—always the military man's prose is characterized by a matter-of-fact tone, with few adjectives, adverbs, or figures of speech—but he did respond to the country in a predictable way. When a landscape calls for grandiose terms, an author will find the appropriate words; when a scene is less majestic, lesser language will do.

Throughout the exploration years of the nineteenth century, Nevada scenery was pictured as anything but beautiful. "A terrific wind blew, threatening for hours to strangle us with thick clouds of sand," writes one of the few women to recount her wagon trip across the desert. "We had now nearly reached the head of Humboldt Lake, which, at this late period in the dry season, was utterly destitute of water, the river having sunk gradually in the sand, until, hereabout it entirely disappeared" (Royce 1932, 39–40). Terse, unemotional, relatively flat—Sarah Royce's diction is as unprovocative as Fremont's, and it typifies those pioneer accounts of the time. Arguably, pioneering was difficult work, and such men and women had little time or energy for the penning of graceful prose. On the other hand, John Wesley Powell must have had just as few leisurely hours and must have been just as scared. The red rock canyon country, more than alkali flats and blackened sage, simply stirs the spirit more emotionally. Writers, male or female, want their wilderness to be beautiful and inspirational. That master of effusive prose, John Muir himself, proves my point.

Not many readers know John Muir's Nevada prose. Originally published in dated newspapers and magazines, collected only once in a hardcover edition long out of print (and not reprinted in the recent rush of Muiriana), the five Nevada essays show off few of Muir's pictorial skills. Even the titles—"Nevada Farms," "Nevada Forests," "Nevada's Timber Belt," "Glacial Phenomena in Nevada," and "Nevada's Dead Towns"—reveal the paucity of Muir's Nevada imagination. A comparison between some of these Great Basin descriptions and Muir's better-known portrayal of the Sierra Nevada demonstrates exactly how "this thirsty land," this "one vast desert, all sage and sand, hopelessly irredeemable now and forever," failed to inspire him (Muir 1918, 154).

Most aficionados of American nature writing are familiar with the more famous passages from *The Mountains of California* (Muir 1977). A long one, found in "A Near View of the High Sierra," Muir draws from a pallet and places in a frame: "one somber cluster of snow-laden peaks with gray pine-fringed granite bosses braided around its base, the whole surging free into the sky from the head of a magnificent valley, whose lofty walls are beveled away on both sides so as to embrace it all without admitting anything not strictly belonging to it. The foreground was now aflame with autumn colors, brown and purple and gold, ripe in the mellow sunshine; contrasting brightly with the deep, cobalt blue of the sky, and the black and gray, and pure, spiritual white of the rocks and glaciers." The paragraph concludes with the young Tuolumne River crossing the scene, "filling the landscape with spiritual animation, and manifesting the grandeur of its sources in every movement and tone" (49–50).

Paragraphs about Nevada, on the other hand, rarely invoke the metaphors of artistry and creation, rarely end on such a spiritual note. "Viewed comprehensively, the entire State seems to be pretty evenly divided into mountain-ranges covered with nut pines and plains covered with sage," writes Muir, "now a swath of pines stretching from north to south, now a swath of sage; the one black, the other gray; one severely level, the other sweeping on complacently over ridge and valley and lofty crowning dome" (1918, 167). Not a single silver state paragraph ends with an invocation, although one does manage to suggest that a covering of snow can make spruce boughs look like a painting. More of the passages, however, are actually denigrating, as when the visitor from abundantly green California "emerges into free sunshine and dead alkaline lake-levels . . . a singularly barren aspect, appearing gray and forbidding and shadeless, like heaps of ashes dumped from the blazing sky" (1918, 164).

Even worse are the ghost towns, scattered "throughout the ranges of the Great Basin waste in the dry wilderness like the bones of cattle that have died of thirst." Muir sees them as "monuments of fraud and ignorance—sins against science" (1918, 203) with no redeeming features. His interest in their decay, however, raises a telling point, for he apparently cared more about Great Basin economics than Nevada aesthetics. Four of his five Nevada pieces—the one about glacial phenomena is the exception—concern themselves with the profitability of the land. "Nevada Forests," for example, concentrates on the "nut pine," and judges that "the value of this species to Nevada is not easily overestimated. It furnishes fuel, charcoal, and timber for the mines, and, together with the enduring juniper, so generally associated with it, supplies the ranches with abundance of firewood and rough fencing" (1918, 169). Subsequent paragraphs pursue first the white man's and then the Indians' use of the "crop."

When Ann Zwinger was thinking about the receding southwestern wilderness, she remarked, "I wonder if the increasing pressures of civilization, of economic exploration, will encroach until the only wilderness we have is that left by default, land that has no value to anyone for anything else for the moment" (1986, 210–211). Her comment, Muir's fascination with meager economic values, and the plethora of canyon essays compared to the paucity of Great Basin ones lead to several conclusions. First—and most obvious—genuine twentieth-century wilderness may indeed be only that landscape which appears to be unprofitable. Now that snowbirds and rafting yuppies have discovered the Southwest, now that the oil companies have found Alaska, spectacular untracked landscape is dwindling and only the economically and/or visually ordinary is left. Second, aesthetics may be a "cash crop," too. A doubter need only look to the national parks for confirmation. Third, if the scenery lacks both so-called conventional "beauty" and financial virtue, a desperate writer will probably hone in on the economics. While it is difficult to aggrandize the former, it is quite possible to conjure up the latter. Certainly Muir managed to focus his Nevada essays that way. Fourth, if neither an aesthetic nor an economic option is available, few writers will concentrate on an area at all. That's why, alas, they don't write about Nevada.

Well, one cannot state absolutely that no one writes about Nevada. But the number of conventional nature or wilderness essayists who have either focused on or noticed the area in passing are few and far between. Even William Kittredge—whose *Owning It All* (1987) proposes to assess "that country of northern Nevada and southeastern Oregon [that] is like an ancient hidden kingdom" (20), and purports to analyze why, when "we owned it all" (60), the dreamland went wrong—cannot sustain his attention on the land. One chapter considers the landscape in depth, but the remainder contain few environmental descriptions and speak more of personalities than places. A critic must not put words into Bill Kittredge's mouth, or condemn him for not writing a different book. The point, though, is that he chose to exclude visual landscape, to include disparate settings even when the drift of the book dictates a kind of unity of place, and not to pen a collection of nature essays. The environment dictated otherwise. He leans toward an economic bias, too, explaining how well-intentioned ranching practices unfortunately wrecked the land. And he especially considers how his stories, retold, affect himself.

If a native son so transparently designs his prose, guests presumably will do the same. The most famous recent literary visitor is John McPhee. His well-known book, *Basin and Range* (1980), documents a recognizably inhospitable Nevada landscape. That he does so in slightly different terms indicates only his style, not his innovation. Since McPhee's narrative voice tries for objectivity, it does not seek self or spirituality in the wilderness. Therefore, his figures of speech are measured, his point of view more photographic than painterly. "This Nevada terrain is not corrugated . . . like a rippled potato chip," he reports. "This is not— in that compressive manner—a ridge-and-valley situation. Each range here is like a warship standing on its own, and the Great Basin is an ocean of loose sediment with these mountain ranges standing in it as if they were members of a fleet without precedent, assembled at Guam to assault Japan" (45).

Even when McPhee considers some abstract qualities of the Great Basin, he holds them at arm's length. "Supreme over all is silence," he philosophizes. Instead of exploring the implications of such a sweeping observation, however, he refers to physicist Freeman Dyson on the subject. Deffeyes, the narrator's companion, finds pleasant "the aromatic sage" (46); the reader never learns what attracts McPhee. Much of the chapter is designed through the use of a distinctive literary device that lends distance—catalogue rhetoric—lists of empty valleys, lists of remote mountain ranges, lists of resident animals and birds, incremental repetitions that themselves replicate the apparent monotony of the basin and range terrain. And how does the central Nevada chapter end? Predictably. "More miles, and there appears ahead of us something like a

Christmas tree alone in the night. It is Winnemucca, there being no other possibility. Neon looks good in Nevada. The tawdriness is refined out of it in so much wide black space. We drive on and on toward the glow of colors. It is still far away and it has not increased in size. We pass nothing" (54). Emptiness and economics, not aesthetics and autobiography, juxtapose.

"We pass nothing." Thus an important American author passes judgment on Nevada. Many readers, I believe, would concur with his assumption, just as most wilderness writers have already agreed tacitly by not turning their eyes toward the landscape of basin and range. Or, when they have deigned to describe the silver state, they have done so in somewhat deprecating fashions—either by focusing on its minimal economic returns or by comparing it unfavorably with its neighbors. John Muir may have found "delightful surprises . . . in the byways and recesses of this sublime wilderness," but at least he knew the sublimity was "scant and rare as compared with the immeasurable exuberance of California" (1918, 164–165).

He would probably have found no reason to favor Nevada over Glen Canyon, either. In direct contrast to what he professed, Muir, and indeed, most Americans, actually prefer wilderness to which human beings can relate. Arches, alcoves, glens and grottoes (see Powell 1987, 230), granite domes, cañons, peaks and valleys (see Muir 1977, 17, for example) are simply more appealing than "a gorgeous, fresh, young, active fault scarp" (McPhee 1980, 49). Conventional aesthetics prevail. Furthermore, the very emptiness that defines wilderness and attracts wilderness writers apparently is the single most unappealing characteristic of Nevada. The irony remains unseen by most essayists.

It would seem that American wilderness writing is following the pattern of our westward movement, civilizing whatever can be touched or tamed while disdaining or ignoring what cannot. Just as settlers made their homes in verdant California long before they filtered permanently into the Great Basin (even today California's population far outnumbers the combined total of its neighboring states to the east), and just as photogenic scenery has always taken precedence over barren wastes, so nature essayists flock to intellectually appreciable locales rather than inhospitable sites. At worst this phenomenon leads to encroachment— whether industrial, economic, excursionary, or literary. At best, it results in some kind of overcrowding and aesthetic destruction. In its own way, then, American wilderness writing is destroying the very places it loves. To repeat, the irony remains unseen.

Glen Canyon and its surroundings have been humanized in prose as surely as in fact. One hundred or one thousand nature essays on a single subject can be almost as stultifying as a concrete hydroelectric edifice; either results in a kind of burial. By comparison, very little humanizing has taken place in the Great Basin. Not many industrial forays have successfully invaded the high desert—although the government keeps trying—and not many authors have been attracted to its desolate power. Thus the area is as much a metaphor for what has not happened in the twentieth century as Glen Canyon is a symbol for what has occurred. Men and women don't write about places like Nevada because traditional modes of appreciation are inappropriate there. Meanwhile, they torment themselves while describing the so-called beauty spots of the West. An unsolvable paradox, perhaps, but one that all wilderness enthusiasts should try to understand.

Notes

1. See, for example, David Douglas, *Wilderness Sojourn: Notes in the Desert Silence* (San Francisco: Harper & Row, 1987), or Rob Schultheis, *The Hidden West: Journeys in the American Outback* (New York: Random House, 1982); even though *The Hidden West* contains essays about inaccessible and less-traveled parts of the West, its best writing occurs in the chapter where the author hikes an obscure, distant, 'beautiful' southwest canyon.

2. When "Why Don't They Write About Nevada?" was published in the Fall 1989 issue of *Western American Literature*, its endnote 19 read as follows:

Two new entries, unavailable before this essay must go to the typesetter, will be published by the time "Why Don't They Write About Nevada?" is in print. Ann Zwinger's *The Mysterious Lands* (New York: E.P. Dutton, 1989) promises to be a welcome addition to the current, necessarily sparse, collection of samples. Stephen Trimble's *The Sagebrush Ocean*, due in fall 1989, from the University of Nevada Press, promises the same.

To add a full discussion of Zwinger's and Trimble's books now, two years after the original publication of "Why Don't They Write About Nevada?" would of course change the flavor of my argument, so I choose to let the original essay stand as it first was printed.

References

Abbey, E. Desert solitaire. New York: Ballantine Books; 1968.

Douglas, D. Wilderness sojourn: Notes in the desert silence. San Francisco: Harper & Row; 1987.

Fremont, J. C. Narratives of exploration and adventure. Nevins, A., ed. New York: Longmans, Green & Company; 1956.

Kittredge, W. Owning it all. St. Paul, MN: Graywolf Press; 1987.

McPhee, J. Basin and range. New York: Farrar, Straus, Giroux; 1980.

Muir, J. Steep trails. Boston: Houghton Mifflin; 1918.

———. The mountains of California. Berkeley, CA: Ten Speed Press; [1894] 1977.

Porter, E. The place no one knew. Brower, D., ed. San Francisco: Sierra Club Books; 1968.

Powell, J. W. The exploration of the Colorado River and its canyons. New York: Penguin Books; [1874] 1987.

Reno-Gazette Journal. Consider scenic Lake Powell if you're looking for a dream vacation. Gannett News Service. 1988 Dec. 11:17c.

Royce, S. A frontier lady. New Haven, CT: Yale University Press; 1932.

Rusho, W. L. Everett Ruess, a vagabond for beauty. Salt Lake City: Peregrine Smith Books; 1983.

Schultheis, R. The hidden west: Journeys in the American outback. New York: Random House; 1982.

Stegner, W. The sound of mountain water. Lincoln: University of Nebraska Press; [1969] 1985.

Thoreau, H. D. Walden. New York: W. W. Norton & Company; [1854] 1966.

Trimble, S. The sagebrush ocean. Reno: University of Nevada Press; 1989.

Zwinger, A. Wind in the rock. Tucson: University of Arizona Press; [1978] 1986.

———. The mysterious lands. New York: E. P. Dutton; 1989.

Original

Wilderness

Prose

Yellowstone in Winter

WILLIAM KITTREDGE

"A G A I N, last night, the coldest place in the nation was West Yellowstone, Montana." That's a story we hear winter after winter on the nightly news, a kind of weatherperson's joke. It's all most American citizens know about Yellowstone in winter.

Which is no doubt a good thing. Winter protects Yellowstone from the hordes of recreational automobilists and bicyclists and tromping backpackers with their multicolored gear, all the cartoon tourists of summertime. You and me and the kids, and Uncle Ted in his Winnebago, and sister Sue whose eyes are blue, everybody in pursuit of a few sweet moments spent checking out a sacred remnant of what we persist in calling wilderness, even after the highways are built. Maybe two million visitors between mid-May and Labor Day.

And then the crowd goes home, leaving Indian summer for those who live in the northern Rockies year-round. Along about the middle of September the leaves go seriously into the business of turning brilliant color along the Firehole River and the other fishing waters. And the trouts, the wonderful slick-bodied trouts, rainbow and brown and cutthroat and brook, are hungry again in the cooling streams and given to lifting ravenously to suck down an ephemeral bit of feather tied handsomely to a barbless hook.

It should be explained that the hooks are barbless if you are an honorable person and only interested in a spot of morning or evening sport. Some meat-eaters go out equipped with a frying pan and a couple of spuds and an onion, and corn flour mixed with salt and pepper in a little baggie. And a lemon. Such persons plan to kill one or more of God's living fish right there on a gravel bar, snap the spine and slit open the tender belly and scatter the guts into the rush for the pleasure of raccoons, and wash the trout body in cold rushing waters, and cook up in the twilight and eat with their fingers.

But much of the time we are not that way. After all, we are the folk who had the good aesthetic sense to live here year-round in the first place. We simply feel it makes good lifetime karma to live in a place where on the evergreen mountain slopes, by early October, the tamarack turns golden, their needles falling through the perfect clarity of afternoon light to litter the undergrowth along the trails like some detritus from heaven.

If that sounds romantic, excuse me. But it's an old custom, our simple-minded xenophobic glorying in the world we have chosen to inhabit in our West. For good reason. Yellowstone is a sacred place, believe it. And it is sacred for reasons beyond landscape and Old Faithful and even the great waterfall on the Yellowstone River. It is sacred for reasons that have nothing to do with our American pride in having at least tried to save some special part of that fresh green continent our people found and overwhelmed with our cities and automobiles and survey lines. It is sacred because of the ecosystem that survives there.

After the tourists are gone Yellowstone belongs to nature again, to the forests and the fungi and, most visibly, to the great animals. All of whom are sacred. It's the way many of us would have it all the time. Fence off the whole damned works, some people say, and lock the gates. God bless them, the extremists, who would kick out everybody but themselves. And I'd be for it if such measures could save the grizzly, who is to my mind the most sacred of all. But that's to my mind.

The damned old gorgeous, terrible grizzly bear. You walk in country where the grizzly lives and you are alert in an ancient way, let me tell you, and in contact with another old animal who walks inside you every day, trapped and trying to get out. Without the grizzly, on our nature walks, we've got nothing to be afraid of but ourselves.

In fall the grizzlies turn irascible as they ever get, absorbed as they are in the hustle for food, layering on fat for the long dozing winter. But we don't mind their outbursts of self-centered crankiness, most of us here, so long as we don't get hurt. What the hell, we say. You don't tear down the Tetons because some climber fell off and got hurt, or killed.

The seasons here are turning, and the thronging thousands of animals are coming down from the high country. The bull elk are responding to hormones and bugling their echoing long cries through the forests, gathering harems and breeding and rigorously defending their lady friends from violation by weaklings and youth. Soon they will be swimming the icy rivers and heading for their winter feeding grounds, mostly outside the Park, maybe 4,000 of them going south to the National Elk Refuge, near Jackson Hole. Another huge herd of perhaps 15,000 will move north of the Park to the country on the west side of the Yellowstone River, beyond the Tom Miner Basin. The bighorn sheep are undergoing the same trials, the rams sniffing the air for signs of females in estrus and running at each other like football players, head to head in heedless combat, the crashing of their collisions echoing among the rock peaks.

The mallards and the Canada geese and all the other migratory water birds gather into vast clamoring flocks and lift into the gray skies of November, heading south. The trumpeter swan does not migrate and endures winter on such open waters as the outlet of Yellowstone Lake. The beaver stockpiles green saplings into his house for winter food, and the timid tiny pika, down there between the rocks on some scree slope, gathers harvest grass to feed on through the winter.

And those few humans who live in Yellowstone during the months between November and March, the Park winterkeepers, they are drying mushrooms and canning peaches and stockpiling three-gallon tubs of ice cream into their freezers. Envy them. Soon the first heavy snows will come, and the Park will be deserted, and they too will settle in, with *War and Peace* or *Atlas Shrugged* or some rugs to weave.

The snows begin to sift through the branches of the Douglas fir, and the interior of the Park is officially shut off to automobile traffic.

Imagine forty below. It's colder than the temperature inside your freezer, and not uncommon when winter has come down on the Park like the hammer it can be. The first heavy snows roll in from the Pacific in great waves, as though they might go on forever in some inexorable end-of-the-world scenario. The plateau around Yellowstone Lake is over 7,700 feet high, and the snow piles up five feet deep on the level. The heavy-browed bison plow along, swinging their heads to sweep away the snow and uncover buried yellow grass. The coyotes prey on tunnel-ing rodents who have come up for air, make predatory moves toward buffalo calves, and study the otters at their fishing, hoping to frighten them away from their catch. It doesn't often work.

The cold is now sometimes terrible, and always there. Wind sculpts the frozen snow, and steam rises from the hot pools. The ice on Yellow-stone Lake sings its music of tension, the coyotes answering back on clear nights. Elk wade in the Firehole, which is fed by hot springs, and feed on the aquatic life. This is winter in the high northern Rockies. Things have always been like this, except for the snowmobiles.

My friend, Dave Smith, who was a winterkeeper in Yellowstone for six years, says the Park in winter is like a woman's body, lovely in its undulations and dappled with secret places.

"Living by yourself," he says, "you make a pact with trouble." Which sounds like a way of saying "death." Some simple mistake, like a bad fall on cross-country skis, can kill you very quickly when the daytime temperatures run to twenty below and the night starts to come in out of the east at four in the afternoon.

"But once you've settled your mind," he says, "then you just go out there, and you find the warm places, where some little steam vent comes up from the thermal. The ground is soft, and green things are growing." Which is what he means, I guess, when he talks about secret places.

"Winter," Smith says, "is a time of dreams."

Burrowing Owls

TERRY TEMPEST WILLIAMS

THERE are those birds you gauge your life by. The burrowing owls five miles from the entrance to the Bear River Migratory Bird Refuge are mine. Sentries. Each year, they alert me to the regularities of the land. In spring, I find them nesting, in summer they forage with their young and by winter they abandon the Refuge for a place more comfortable.

Our relationship is generational. They know my relatives and I know theirs. At least, that's the myth I choose to believe. We first became acquainted in 1960, the same year my grandmother gave me my Peterson's Field Guide. I know because I dated their picture.

What is distinctive about these owls is their home. It rises from the alkaline flats like a clay-covered fist. If you were to peek inside the tightly clenched fingers, you would find a dark-holed entrance.

"*Tttss!*" "*Tttss!*" "*Tttss!*"

That is no rattlesnake. Those are the distress cries from the burrowing owl's young.

Adult burrowing owls will stand on top of the mound with their prey before them, usually small rodents, birds, or insects. The entrance is littered with bones, feathers and fur. I recall finding a swatch of yellow feathers like a doormat across their entrance. Meadowlark, maybe.

These small owls are able predators pursuing their prey religiously at dusk.

Burrowing owls are part of the prairie community, taking advantage of the abandoned burrows of prairie dogs. Historically, bison would move across the American Plains, followed by prairie dog towns which would aerate the soil after the weight of stampeding hooves. Black-footed ferrets, rattlesnakes, and burrowing owls inhabited the edges, finding an abundant food source in the communal rodents.

The protective hissing of baby burrowing owls is an adaptive memory of their close association with prairie rattlers. Snake or owl? Who wants to risk finding out?

With the loss of desert lands, a decline in prairie dog populations is inevitable. And so go the ferret and burrowing owl. Rattlesnakes are more adaptable.

In Utah, prairie dogs and black-footed ferrets are endangered species, with ferrets almost extinct. The burrowing owl is defined as threatened, a political step away from endangered status. With development in Utah on a rampage, burrowing owl habitat is being plowed under with agricultural interests and subdivisions.

Each year, the burrowing owls at Bear River become more blessed.

Not many people knew about the owls and I rarely told. Most visitors to the Refuge didn't even look for birds until they crossed the bridge to Headquarters. But you learn birds are not respecters of boundaries, especially ours. They define their own. And the owls had staked their territory just beyond one of the bends in Bear River.

I could see their silhouettes a half-mile away, bobbing up and down on top of the mound. Just under a foot, they have a body of feathers the color of wheat, balanced on two long, spindly legs. They can burn grasses with their stare. Yellow eyes magnifying light.

Whenever I drove to the Bird Refuge, I stopped at their place first and sat on the edge of the road and watched. They would fly around me, their wingspan sometimes two feet long. Undulating from post to post, they would distract me from their nest. But from the corner of my eye, I could see the young trickle out of the burrow; one, two, three, four, five . . . seven was not unusual, although by the end of the season, two or three was a more likely head count.

In the twenty-five years I grew accustomed to their ways; I'm not certain they ever accepted mine. They couldn't afford to.

While working as a naturalist in Grand Teton National Park, I met a lovely couple impassioned with birds. They were particularly fond of owls. Both of them were expert mimics as we sat on an old log in the lodgepoles and hooted. They were traveling west from the east and had planned a stop at Bear River. I judged them to be of secret-keeping character and told them about the owls. They promised to tell no one. I sent them with my blessings. A few weeks later, I received a 5" x 7" color photograph of the two burrowing owls, lemon-eyed and healthy, standing in front of their mound. On the back of the print it read, "Greetings from home."

A fellow employee asked me who the picture was of and with tongue in cheek I replied, "My family, would you like to see?"

"Never mind," he said, walking off. "They all look the same."

That was his mistake and my pleasure.

In 1983, I worried about the burrowing owls, wondering if the rising waters of Great Salt Lake had flooded their home, too. I was relieved to find not only their mound graciously intact, but four owlets standing on its threshold. One of the Refuge managers stopped on the road and commented on what a good year it had been for them.

"Good news," I answered. "The lake didn't take everything."

It was late August and the sky was unusually still. A few shorebirds were feeding between the submerged shadscale. Through my binoculars I could see long-billed dowitchers and godwits.

In November, a friend from Oregon came to visit. We had spoken of the Bird Refuge many times. The whistling swans had arrived and it seemed like a perfect opportunity for a shared day at the marsh.

To drive to the Bear River Migratory Bird Refuge from Salt Lake City takes a little over one hour. I have discovered the conversation that finds itself into the car usually manifests itself later on the land.

We spoke about rage. Of women and landscape. How females and the environment have been objects of exploitation.

Intellectually, I supported the argument, but personally I was not convinced, finding the rhetoric of angry feminists too simplistic, too hostile, too exclusionary.

"It's not a matter of rhetoric," my friend explained. "It's a matter of experience. Perhaps your generation, one behind mine, is a step removed from the pain."

I argued again from my own perspective. "It's just that I personally don't feel rage. I feel sadness. I feel powerless at times. But, I'm not certain of what rage really means."

The conversation shifted to women writers and poets, both of us recounting favorite stories and stanzas.

We reached the access road to the Refuge and both took out our binoculars ready for the birds. Most of the waterfowl had migrated, but a few ruddy ducks, redheads, and shovellers remained. The marsh glistened like a cut topaz.

As we turned west towards the Refuge, a mile or so from the burrowing owls' mound, I began to speak to them, *Speotyto cunicularia*, how they were my continuum with the Refuge. I recounted the time when my grandmother and I first discovered them, hardly believing our eyes, how we had come back every year since to pay our respects. I explained the goodness of the year, how four owlets had survived the flood. We anticipated them.

About a half mile away, a peculiar knot began to tie itself in my stomach. I could not see the mound. I took my foot off the gas and coasted. It was as though I was in unfamiliar country.

The mound was gone. Erased. In its place, fifty feet back, stood a cinder block building with a sign, CANADA GOOSE GUN CLUB. A new fence crushed the grasses with a note posted, KEEP OUT.

I got out of the car and walked to where the mound had stood for as long as I had a memory. Gone. Not a pellet to be found.

Just then a blue pickup pulled alongside us.

"Howdy." They tipped their ball caps. "What y'all lookin' for?"

I said nothing. My friend said nothing.

"We didn't kill 'em. Those boys from the highway department came and graveled the place. Two bits, they did it. I mean, you gotta admit those ground owls are messy little bastards. They'll shit all over hell if ya let 'em. Our winda sills were a sight fer fools. And try and sleep with 'em hollerin' at ya all night long. They had to go. Anyways, we got bets they'll pop up someplace around here next year."

The three men in the front seat looked at us, then tipped their ball caps. "Well, no point stayin' here." And drove off.

Restraint is the steel partition between a rational mind and a violent one. I knew rage. It is the white plaster mask with the searing eyes behind it. It was fire in my stomach with no place to go.

A few weeks later, I drove out to the Refuge again. I suppose I wanted to see the mound back in place with the family of owls bobbing on top. Of course, they were not.

I sat on the gravel and threw stones.

By chance, the same blue pickup with the same three men pulled alongside. It was obvious they were the self-appointed proprietors of the newly erected Canada Goose Gun Club. Employees from Thiokol.

"Howdy, ma'am. Still lookin' for them owls, or was it sparrows?"

One winked.

I suddenly pictured the burrowing owls' mound perfectly, that clay-covered fist rising from the alkaline flats. The exact one these beer-gut-over-belt-buckled men had leveled.

I walked quietly over to their truck and leaned my stomach against their door. I held up my fist a few inches from the driver's face and slowly lifted my middle finger to the sky.

"This is for you—from the owls and me."

[Editor's note: The black-footed ferret (*Mustela nigripes*) is believed to be extinct in Utah.]

The Arctic Desert

HOWARD McCORD

There is absurdity here and there is
darkness, but they belong to our conceptual
systems and not to existence or the soul.
S.M. ENGEL

I MAKE these notes hurriedly, copying some from three note-books I carried while traveling, and from what memory has left me.

Light is the dominant element: midnight is gold, creamy. Noon is silver. I forgot about the stars, and was astonished by the moon when I saw it again, two months later. Light does away with stars, which once seemed such treasures. I never noticed the sun—it wandered ceaselessly, quietly. Perhaps what I felt is linked to a phrase I found in myself watching the miles of the Springisandur before me: a space on which a kind of eternity rests. Stars define time, but light removes it. Erases it even from the body. Slips it off, and you walk without a second's shadow darkening your eyes.

So much earth and sky are accessible to vision in the interior. My eyes were filled, welling with immensities and brilliance. I was glad for the simplicity of the landscape, for to perceive a city or a forest with that clarity would be an annihilation, senses drowned. It was curious to observe my eyes. Where are things located? What is out there? Music not in the score, a poem not in the words? What is in the landscape? What is in my body? All the minima are observed, a gathering of distances.

But there was no awe. Curiously no awe.

What we see is how we look. Even asleep, I could feel the light pressing through the tent, pressing my body. Before I came to this place I defined a desert as "an excess of simplicity." Now I am sure of neither the key words. I would take the word "virtue" in its old alchemical, medical use. Then the desert shifts. A note: "Solitude, as its opposite, is both positive and negative, healthful and destructive. Then, is it the field, the context (in all ramifications) and the motion of the constituents of the field which generate the Virtue?"

What is the virtue of desert, and what are the correspondences to be maintained? From the Landnámábok (38) "the farm stood where the mountain is now." Mountains ignore time in Iceland, too. They grow like the sky changes. I come to my own rhythms, and must stick by them. I walk at such a pace, no other. Simplicity allows discovery. There is a linkage in the paradigms of clarity, isolation, and freedom.

I sit by a second-story window overlooking Main Street. The traffic is relentless in the dark. A few weeks ago I was back at center, looking out toward Arnarfell from a ridge above the Jökuldalur—31 kilometers by line of sight. It looked like two hours' walk. From a week before on a northern cape, at Breithavik, I had seen Grimsey, squatting on the Arctic Circle, its cliff distinct at 52 kilometers. The white walls of the house across the street flicker like moths' wings in the passing headlights. I am enclosed.

"A space on which a kind of eternity has focused, and lingered." And the processes of the body, are they caught too? What do we do when we do nothing? Master Blake tells us, "Time is the mercy of Eternity," and so it is. Perhaps I should not even ask, and accept the mercy of so much space. Here on this stairway landing, I have surrounded myself with sacred objects. Some have power on their own account, others possess it as an adjunct to friendship, or to a simple wish. A kachina, a magical egg, knives, a feather pajo I made of a summer's feathers—feathers the god wind brought to my feet. Baby Krishna with his butter pat, the great mother of the Cyclades, a Tibetan charm, one-half a stone map, a wooden plaque telling me "Haz todo con amor." Dozens more in the desk drawers, each with a complete story. A thorough accounting would produce an autobiography, for at each turn I have taken, I have picked up the physical marker, the marker given. I burned twenty years of letters last May: everything in them is contained in six objects. I took no charm to Iceland, wore no rings to fill with greater light. I brought back a small piece of new lava from Askja, needle-sharp, drawn out and shaped by a dynamic I do not understand.

This night I feel these objects wail and grind a singular "I" at me, as though they were parts of my body, as they probably are. But that I was freed from in Iceland. I spoke little, I wrote little, I needed no books, no objects such as these to consolidate experience, to stand as substantial symbol and code, my history compressed into matter. Silence and light were sufficient. To celebrate them is to become invisible, even to oneself. I was a deer once, in New Mexico, through Tío Híkuri. In Iceland, I had not even a shadow's weight, doing nothing, the distances swallowed in light, and emptiness filling my eyes with a sensation so pure it required no perceiver to exist. White noise, totally, white mind.

I love the genial stupor of long walking, the slow dissolve into foot-steps, into one's boots. All that exist are balance, footing, and if the land moves up, thrust. My eyes are 163 centimeters from my toes—I look with my feet, walk with my eyes. In the lava fields a golden plover will flutter to a stone, call you away from his mate, follow you a mile, talking, deceiving you.

On the fifth day walking, a bird screamed high in the air, from the heights of Singitjakka, an eagle, the first I've ever heard. The cliffs rise a thousand meters there. He was a black speck, circling at the rim. I watched a bit, then put my mind back in plod-time, measuring the rhythms of Nothing.

In the black desert, tiny markers of pink, *Armeria*, called thrift, mixed with moss campion, the dwarf catchfly, *Silena acaulis*. I cannot imagine what creature pollinates them. There must be a species of insane Icelandic bees—tiny, fearless, with a bizarre metabolism. (But it is the wind.) Gnats and midges at Herthubreitharlindir, and wonderful black orb-weavers in the lava, feasting on the gnats. At one campsite in the lowlands, I found mosquitos dormant, sleeping quietly in the petals of roseroot, *Sedum rosea*, too cold to fly. That was at Thorsmork, in the south.

The notes, continued. Begins with one word in Icelandic: ÓBYGGDIR, interior. In the sagas, VESTRI OBYGD was the western wilderness, perhaps Baffin Island. The Norse Greenlanders called the harsh east coast of Greenland UBYGDER, the "unbuilt" places. No homesteads were found east of Cape Farewell, and the tales run to shipwreck, and separation. Torgils Orrabeinfostre spent four years making his way to the western settlements after a wreck in 1001. A hard journey, perhaps not even true. But hard.

And I am back in New York City. Walk the street, and flashes of recognition. I think I see acquaintances, hear my name spoken, the shuttling crowd like a loom at work. The physical feeling is a déjà-vu fatigue. Then the weight of the crowds presses down. I look at *American Poetry Review*, and it is an incessant scream to run, run, run, write faster. These people write. They must do it every day, as a sickness, an expiation, a sacrifice demanded by loud noises, movements.

I come to my own rhythms, and must stick by them. There is a madness in this City that even Fraenkel did not grasp. I go see HEAVY TRAFFIC. It is perfect. Inside and outside the theater are finally one. The topological problem of turning the inner tube inside out is experienced, is. The crowd is anxious, bored. Two in the theater sitting next to me drink from a pint flask, and then make acute philosophical observations about the film. Carefully, I look at them. Two young men I would take to be stockclerks, GIs, construction workers. One explains to the other the sociological significance of the work observed. When I leave I step over the seat and go behind them. I have no wish to get into a fight.

At the Algonquin, George behind the bar tells me that the Japanese are everywhere, with cameras, as before the world war. I should buy Ashland Oil, because something is brewing. It is the Chinese who are using the Japanese, as the Germans used the Japanese before World War II. The Chinese will soon have us. He has a son who gets into trouble and then out of it. Wrecks cars regularly, drinks too much. I like George very much. In Narvik, the French boys I roomed with in the hostel told me when I came back from a climb, "The children here, they are all drunk!" They were indignant in their surprise. I would be drunk most of the time if I lived in Narvik too. But not in Iceland. The kids had nothing to do but roar down Kongens Gate on their Kawasakis, drink beer at $1.40 a glass, and rut with local girls, who were at least as bored, and maybe more, with no Kawasakis. I left Narvik the next morning and began my Lapland walk without equipment I might have purchased there. No regrets. George has an excellent memory. We went over a minor occurrence two years previously. He remembers. The man involved has now been put away, somewhere in the Carolinas. It has to do with a birthday cake and the CIA.

I realize the strange independence required, the resolute cantankerousness I must employ. I never felt this in Iceland. "Argue from Phenomena, and do not feign hypotheses," says Newton. OK. OK. In the city, the "something wrong" has a tangibility it lacked before. My fingers do not curl on a glass in quite the same way. I can no longer be abstract. At 11 P.M. I am in my room, and hear police whistles. Many, many. I expect to hear gunfire. I do not. The next morning I realize that it was the doorman calling cabs. All night I think about Nothing. I play Wittgenstein: "What do we do when we do nothing?" What is the difference between the minima of the Desert, and Nothing? Is minima divisible? Can one say, of line AB, there is first A, then B?"

God has no memory, only perception.

At home I read Guy Davenport on Agassiz, "The Western world has had three students of metamorphosis. Ovid, Darwin, Picasso." I am assuming he wrote this before Portmann. But it fits the City. In the desert, one does not make up scenarios for the stones; in a New York bar, it is nearly impossible not to. Everyman the dramatist of his proper moments. Across the small room, six people argue analytically. It is worse than high school and the heady discovery of logic. At the bar, people dance slowly, standing still. But their talk is movement. Control invites tension. One gentleman's words lick the nipples of the woman beside him, who is talking to another gentleman. There is a contest everyone in the room is playing—Who Is the Best Fuck of All? Alone, in an odd corner, I am a player too. All must play. No coercion. Simple presence is membership. The freer people get, the more they drink, the tighter grows the game. I drink to keep up with the game. I have spoken to no one. George is busy at the bar. He does not need to talk about conspiracies when he can be a part of one. The lady in the alcove is the daughter of a famous radio and film writer, recently retired. She is small, red haired, and boasts only a little. I have no desire to meet anyone present. I remember a line from the Landnámábok again: "They left Telemark because of some killings" (6).

You see, the desert presents a task. New York City presents a burden. I hear a sentence from the bar: "He is careful about his company." So it is here that I should enter Boswell's sentence from his CORSICA: "Where they make a desert, they call it peace." He found the line in Tacitus.

The peace I find resides in imagination, in the gentle whirling of my mind to the rhythm of walking, the loose play of words jostled against one another, jostled enough to make a sweet confusion. . . . It is the very premise of a notebook.

There are two geographic fantasies: "the Antarctic Convergence" and "the Pole of Relative Inaccessibility." I love the latter better.

Take, in your mind, a likeness of this earth. Detach the southern hemisphere from the northern. Make a montage of S and N. Notice where Antarctica extends. The northern tip of Iceland is just about where the south magnetic pole is. 66° 33' North or South. A curious number. It measures a twenty-four-hour day, the extremes of. Ask someone for a definition of the Arctic or Antarctic circles.

The wind circles the earth from the west, in the south. That's Coriolis. Do you know which way the Arctic Ocean rotates?

In Iceland, the sky is not an acquaintance, I hold no token of friendship. I would not forget something which so surely could not be given.

Now, the Pole of Relative Inaccessibility. 84° North, 160° West. It is the most difficult region to reach by surface travel. A refinement of the old "Ice Pole," marking the center of the polar ice cap (that was held to be 86° N, 156° W), it was nearly attained by Wally Herbert and his friends as they made their way by dogsled from Point Barrow to Spitsbergen in 1968. I have never been there, am not even interested in going. I would love to walk Kamchatka, or winter in the Jarbidge. I will tell you one day's travel, to Hveravellir.

The forty-eight-hour tourist in Iceland often takes a little trip out to Thingvellir, where the Althing was founded in 930 A.D. Then they sometimes go farther, east through green Laugarvatn, to Geysir, and then on to see the Hvitá roar over the Gullfoss. But there they stop. They turn around. The road stops, and the track begins, north. The earth turns dark grey. To the west are the cones of Skjaldbreidur and Höldufell, one a shield volcano, the other tight, steep, assertive. The road must stay to the west of the Hvita, much too powerful a river to ford. The land has gradually risen to 500 meters, and the track points to a pass between the huge Langjökull glacier and snow-topped Bláfell, 1207 meters. We stop at the pass, throw a stone on the troll-pile, and look down on Hvitárvatn, motherlake of the Hvita, fed by a tongue of ice from the glacier, Nordurjökull. Where the Hvita begins, a construction crew repairs a tiny bridge. We cross on foot, the driver takes the Bedford across alone. We will meet no one else for more than a hundred kilometers. Loury winds beat up from the south, spill down from Langjökull. The track leads through the Kjalhraun, an old lava field—bleak, dark, filled with twisted, startled rock. At one point, the Eyfirdingavegur branches off, noted on the map as an Indistinct Path. It goes to the NE, around Hofsjökull, the glacier to the right. Hofsjökull, like Langjökull, covers more than 2000 square kilometers. It is a trail for madmen. And it brings Eyvindur into the account.

The road swings to the west, climbs, drops a bit, and steam rises in the sky from a dozen hot springs. This is Hveravellir. The wind is cold and spitting rain. We pitch our tents, eat, then, at midnight, climb into a spring to soak. Some yards away is another spring, boiling, where Eyvindur cooked his mutton. He spent a year or so here, hiding, wandering, of all his years in the interior, wintering where nothing but lichens had wintered before. He and his wife, Halla, spent twenty years in outlawry, útilega, life in the open. Two hundred years ago he built the walls that hold this spring's waters. Rebuilt since, I thank him, my head sticking out of the water, nose freezing, body hot and sleepy. In the bad winter months it must be a miracle of warmth. We run back to the tent, after dressing in the 0° C wind. Scotch, then sleep.

One day's travel.

The spoor of Eyvindur is everywhere in the highlands—at Herthubreith, in the Sprengisandur, in the lost valley, Thórisdalur. North of Hofjökull a huge area—5000 square kilometers—is named for him, Eyvindarstadaheidi, as empty of trails today as it was in his age.

Johann Sigurjónsson, who turned the tale he read in Jón Árnason's *Thjóthsögur og Aefintýri* (Leipzig, 1864–74, 2:243–251), into an immensely popular Icelandic play, wrote to his friend Pineau: "Pour pouvoir l'ecrire en toute véridicité, j'ai parcouru à pied cinquante milles du nord au sud de l'Islande, donte trente à travers le desert au milieu des glaciers où ejvind et sa femme ont vécu." Here is part of his story, adapted from Magoun's Englishing in *PMLA* (1946, 269–92):

He is said to have pilfered cheese from a beggar woman's bag and to have been at that time in Oddgeirsholar. She laid a spell on him that he should never stop stealing from then on. Then either Eyvindur or his people wanted to bribe the old woman into taking back her words; she said she couldn't do that, because words mightn't be taken back, but that she would remedy the matter to this extent, that he would never fall into the hands of the law. This point seemed to prove true ever afterwards in his case.

In many respects the wife Halla seems to have been a poor lot: she was of harsh disposition, had a bad reputation, and was thought to be unorthodox in her faith so that she scarcely went to church or stood outside the church door while the service was going on. Her physical appearance and manner were described at the Parliament (at Thing-vellir) in 1765: she was "short and lordotic, her face and hands very dark, hazel eyes and heavy eyebrows, adenoidal, long-faced, very ugly and ungainly, dark haired, with small thin hands, used a great deal of tobacco."

Eyvindur, on the other hand, was apparently very well endowed, of good and cheerful disposition, athletic, a good swimmer and glíma-wrestler, very swift of foot, and an excellent climber, so good at turning cartwheels that he out-sped the swiftest horses. That often turned to his advantage when he was pursued and needed to save his life.

Eyvindur built himself a hut, and traces of it are still visible west of the Sprengisandur Route. The roofless walls of the hut have now almost entirely collapsed, but a spring runs out of it in three directions. The stream that runs out to the northwest is full of horse bones which had been chopped up for meat, and some bones of birds; sheep bones have been found there. Eyvindur and Halla are said to have lived here the greater part of their outlaw life.

One Sunday in the summer when divine service was being held in Reykjaklith—the church stands somewhat away from the farm and is surrounded in every direction by lava-fields—Eyvindur asked to be allowed to attend services. He appeared to be a devout person; Halla paid no attention to the business. Permission was granted. Eyvindur sat down in a pew near the door, and they thought it wouldn't be necessary to keep watch over him during the service; otherwise two men generally watched him. But while the minister was intoning the Gospel and every-body had their eyes on the latter and no one was looking at Eyvindur, he disappeared out of the church and wasn't searched for until the service was ended. But then a pitch-black fog came up suddenly, so that you could hardly tell one man from another. This fog lasted day in day out for a week. Since that time Myvatn people call every pitch-black fog an "Eyvindur-fog."

A long search was made for Eyvindur and nothing came of it, but, as he himself afterward related, he hid in the lava-ridge next to the church while the search was at its height. Nobody thought of that, and they looked for a long time for what was right near them. The winter after he vanished from Reykjahlith, Eyvindur lived at Herthubreitharlindir and traces of his hut there are still to be seen. It is a stone enclosure, built up against the wall of the lava flow, approximately a good 6" x 3" feet. He had a horse's spine as a ridge-pole in the hut, and a willow branch was pulled clear through the spine to hold it together; afterward it was thatched over with a layer of lyme-grass roots. In the doorway was a slab of stone as well fitted as if it were planned. A spring gushed out of the rock against which the hut was built and flowed down right past the bunk of the occupant. The spring was so skillfully contrived that, from his bunk one only had to reach out, lift up a stone slab that covered the spring, and lower a vessel into it. A big heap of dry roots and branches was near the hut, and people think that Eyvindur kept his winter supplies in it. Eyvindur is reported to have said that that was the very worst winter he had while outlawed; there was nothing to eat but raw horse-meat and angelica roots, of which there is plenty at Herthubreitharlindir. Eyvindur is said to have stolen seven to nine horses from the Möthrudalsfjöll in the autumn, but there were a few sheep to be had in the vicinity.

After Eyvindur got back to the settlements he said that nowhere had he been better off as an outlaw than at Eyvindarver; for aside from the fact that he took sheep from the grazing lands, he had had lots of swans and geese there, running them down when they were moulting. Furthermore, he was able to avail himself there of the trout fishing which is said to be inexhaustible in Veithivotn and Tungna-á, though these lakes were rather far off. However, Eyvindur said that the freezing cold winds on Sprengisandur were sometimes so severe that a man in his full vigor and well clad couldn't survive out of doors. Therefore it is more likely he said he wished no one so ill as to be able to wish him his life, that what he is also supposed to have said, namely, that he had no enemy so hated that he would want to direct him to the western desert, but that he would be willing to direct a friend to the eastern desert, east of Odathahraun.

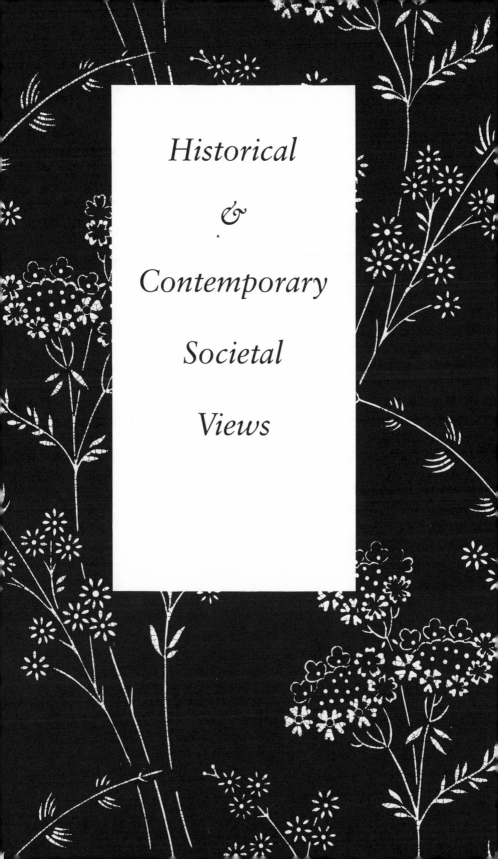

Historical

&

Contemporary

Societal

Views

The Bob

Confessions of a Fast-Talking
Urban Wilderness Advocate

MICHAEL P. COHEN

My own belief, which I realize the majority do not share,
is that most exploration today is not of material value to the human race in
general but is of immense value to the person who does it.

BOB MARSHALL

Maybe in the days of the pioneers
a man could go his own way,
but today you gotta play ball.

BURT LANCASTER[1]

If you ever forget you're a Jew,
a gentile will remind you.

BERNARD MALAMUD[2]

Well, no cowboy was ever a Jew.

JOSEPH HELLER

Vice Versa.

MICHAEL P. COHEN[3]

W E   W I L L know him first of all by his appearance, his shaggy hair, rumpled flannel shirt, khaki pants, wool socks, and sneakers (Glover 1986).[4] He is a homely looking man with prominent ears and protruding eyes (Fox 1981). An ungainly looking man, we are told (Meine 1988). Look at the pictures. It is true. Here is a man with bulging eyes and an open expression, with an awkward boyishness and a silly grin. This is not the Montana Face (Stegner 1987). Even as an adult, he posed like a child, and one notices that he is always on the edge of a group. Here he is in Idaho; everyone has on a western hat but Bob (Glover 1986). [One wonders whether his lack of hat is associated with the sunstroke he later claimed to have had, just before he died; was this a joke? (Glover 1986).]

Never was there a less satisfactory Westerner. Afraid of the dark (perhaps afraid to be alone), son of a protective mother—a mamma's boy—The Bob always demonstrated a lack of finesse with tools, including a canoe paddle. He was a terrible driver of autos, a horrid, though creative, cook (Glover 1986). Surely he was one of the worst outdoorsmen ever to arrive in Lewis and Clark country.[5]

It is not easy for the grandson of an immigrant and the son of an urban Jew to adapt to the code of the West. As Hannah Arendt (1978) has noted, Jews have never liked to call themselves refugees, but have preferred the terms "newcomer," or "immigrant." Further, Arendt notes, one escape from the status of nonentity which Jews have suffered as a pariah people has been to the world of nature in which all men were equal under the sun, or to the realm of art, where taste was the only prerequisite.[6]

No wonder that, as a Forest Service employee, The Bob failed on several grounds. First and foremost, he was a slob when it came to housekeeping (Glover 1986). He had eternal problems organizing his life and reducing the chaos of his work. As his supervisor wrote, he was "very weak in the mechanics of field work and field living . . . awkward with tools and equipment that are necessary to a forester's work . . . and you lack orderliness in certain aspects of field work" (Glover 1986, 61).

It gets worse. He wore no proper uniform, and when the yellow jackets would get into his pants, he could only shriek, "Oh dear, oh dear. . . ." As a fellow forester said, "If you have to cuss for God's sake cuss like a man" (Glover 1986, 57).[7] Unable to cuss properly, and unfortunately not closemouthed either, Bob Marshall was, by all accounts, never prone to understatement (Glover 1986).[8] Though he may have been a slob as a housekeeper, he was fastidious about conversation. Though he could be sympathetic with the human misery of laborers, he was displeased with what he called the "smutty talk and grumbling of fire fighters" (Glover 1986, 71).

This man was no fighter at all; certainly no brawler. He never drank, so he could hardly become a bar brawler. He preferred the use of tact and firmness, the use of disarming humor (Glover 1986).[9] It was not always gentle humor.

Of course he was eccentric. He was an obsessive rater of almost every-thing (and he in turn has been rated) (Glover 1986). Baseball had been his childhood game with its silly statistics, and something more. He was an obsessive lone walker with the frivolous ambition to do a thirty-mile hike in every state. He spoke of this as following "a rather foolish, but interesting ambition" (Glover 1986, 56, 58). Whether this was com-pulsive peak-bagging, represented inner conflicts of personality, or the natural pleasure of so-called "peak experiences," Bob Marshall was always happiest afoot (Glover 1986). He loved square dancing, but was, according to one observer, "a real funny dancer" (Glover 1986, 178).[10] As for horses, he was not much good with them, and as a man who kept his feet on the ground, he naturally didn't like to ride them (Glover 1986).[11] It is quite possible that this is the basis for his later, and more well-known statement when creating The Wilderness Society, "We want no straddlers" (Fox 1984, 5). What can we say about the turn of mind of a man who writes an article correlating precipitation and presidents, whose two favorite pastimes are walking and people watching? Perhaps that he is the Nathanael West (also known as Nathan Weinstein) of the environmental movement (Glover 1986, 62, 65).

This is what Bob Marshall's life might mean: he is the picaresque hero who invents himself. He is a comic hero we need because he provides an antidote to all the humorless and moralistic stone-faced wilderness advocates we see so frequently. Marshall's consistent good moods and constant enthusiasm remind us that he is not "tragic man."[12] He is my true and authentic hero.[13]

The view of The Bob as hero was not Richard McArdle's version of The Bob. You know McArdle, who became Forest Service Chief in 1952. He was the one who tried to eat The Bob's famous "Broiled Eggs" back in 1924. (Marshall announced that broiled eggs were going to be the national dish. McArdle didn't laugh. In true Scottish style, he said they tasted like golf balls.) In 1952, McArdle inherited all of The Bob's maps of wildernesses. McArdle thought they were sloppy maps, because they enclosed too much; they were, he said, "free, wide, and handsome." What was worse, by this time, Marshall "had become some sort of holy guru to the wilderness people" (McArdle 1975, 178–179). Marshall wasn't sloppy with his boundaries, he was generous (as he was in all things). Generosity—even "reckless magnanimity" (see Arendt

1978, 67)—that quality missing from the developing American character as depicted by Frederick Jackson Turner. McArdle was niggling and penny pinching; he didn't know what generosity might be.[14]

We inherit Bob Marshall, or rather we have been inheriting him through various versions of The Bob, beginning with Nash's "The Strenuous Life of Bob Marshall."[15] Nash provides us with the deracinated Bob, where the man's chief virtue was not originality, but rather zeal. Nash declared that Marshall was a "case in point" of "the tendency of wilderness enthusiasts to arise from refined, urban situations" (Nash 1967, 201–203). Never once does he mention that Marshall was a Jew. Then one has Stephen Fox's Bob, a "fuller" embodiment of the Sierra Club tradition, capable of being compared to Muir—indeed Fox moves easily from Muir's "Lord Man" to Marshall's "the puniness of man." Fox's Bob has a distaste for modern life; his socialism contained within an "unabashed elitism," he is a man of internal conflict whose interests are "neatly compartmentalized into socialism, civil liberties, and conservation" (Fox 1981, 206–208).[16] Kurt Meine (1988) has, by contrasting Marshall with Leopold, shown the extent to which Marshall, unlike Leopold, was a political man, and in a way which would someday cause problems for The Wilderness Society.

Most recently we have Glover's colorful, three dimensional, reportorial biography. In his assessment of Marshall's role in the wilderness movement Glover finds Marshall not so eloquent as Muir, Thoreau, or Leopold, Marshall's arguments not as ecologically sophisticated as Leopold's or Murie's. Which is to say that Marshall has been overshadowed by his more literary colleagues (see Shands 1987). The contributions of environmental historians to our understanding of Marshall's importance have been additive, but not cumulative. Something is missing (Glover 1986).[17]

It is precisely the clashing of urgency and rationality (with a sense of the absurd) that constitutes the style of The Bob. He writes, "Within the next few years the fate of the wilderness must be decided. This is a problem to be settled by deliberate rationality and not by personal prejudice" (Marshall 1930). Here is Marshall's Jewish style. It is a literary style. (Do I contradict myself? Do I contain multitudes?)[18]

What if it turns out that The Essential Bob is the Eccentric Bob? (Or vice versa.) Why not? To begin with, one might want to start with the importance of Jewish ineptitude, a common problem usually attributed by Jews to the sons of rabbis. This phenomenon has been vastly under-estimated, and is without question a formative influence on Marshall's wilderness experiences. All a man like Bob Marshall had to do to gain proper humility was to try to be a man while playing by the code of the West. Such a course led naturally to a nonanthropocentric perspective.[19]

You can tell a lot about a man by the kinds of stories he approves and disapproves, and by the stories he tells about himself. Early in his west-ern experience Bob recognized the vanity of lumberjack stories, which was, at bottom, linguistic. When not pornographic, the lumberjack's story was most likely to consist of "personal adventures in which narra-tor is hero" (Glover 1986, 92–93). On the other hand, humiliation was the common theme in Bob's wilderness narratives, and he could laugh at himself.

One day in Montana, Marshall was treed by a sow grizzly; unfortu-nately gravity did its worst, he fell out of his perch, and he had to use the more troubling ploy of playing dead. The ploy worked, but, as he said, "It was a terrible blow to my self-esteem" (Glover 1986, 86). Mar-shall considered the event worthy of a humorous telegram home to New York. Once he cited with approval the Indian suggestion that grizzlies might be introduced to the Big Horn Canyon "for the partial purpose of scaring out white intruders" (Glover 1986, 210). Another time, while his camp was being swept away by an Alaskan river changing channels in the night, Marshall had a dual consciousness—one responded to the emergency and the other watched and celebrated that nature had not been conquered (Glover 1986).[20]

Actually, Marshall suffered many happy disasters in his explorations in Alaska. There was a near fatal boat wreck on the Koyukuk River, where Marshall thought he might die. The journey itself seemed, to The Bob, "without any anti-social byproducts" (Glover 1986, 242–244). Better still was the ascent to the Arctic Divide. When it turned out not to be the Arctic Divide, Marshall named it Kinnorutin, meaning "You are crazy" (Glover 1986, 136–137). The best was the failure to climb Mt. Apoon, for reasons of technical difficulty. Marshall could be happy under the circumstances, having done his best, when he recalled:

I had talked only recently with an assistant manager, an associate profes-sor, and a division chief whose lives for several years had been unhappy because they had not been promoted to head manager, full professor, and bureau chief. (Glover 1986, 257)

What could be clearer? The consistent message is that Marshall did not go to the wilderness for conquest. In many ways, he realized, it was a better experience when one did not return a hero, and when one learned humility. In this sense, The Bob's wilderness excursions were *not* failures.[21]

The man who did not go to the wilderness to conquer Nature was not likely to become a "charismatic" leader, thank God.[22] Do we need more "charismatic" leaders? No, we have plenty of those, more than we need. We need someone who was always political, a member of the ACLU like The Bob, not a "compromiser," but a man who yearned to be "Man-with-the-ability-to-bring-all-factions-together" (Glover 1986, 170). Bob Marshall was a social man; he was—please excuse the term— a facilitator.

Indeed, The Bob was not a charismatic leader, but quite the opposite; he was our first environmental deconstructionist, the first wilderness advocate to go through modernism, to break through the wall of pessimism, and come out the other side, shaken, cold, humiliated, but smiling. Of course he read Spengler, the early Joseph Wood Krutch, Bertrand Russell, and Aldous Huxley (Glover 1986). He understood them well, yet retained his idealism while being able to laugh at it. When The Bob asserted that the essentials of mechanized society were drabness, nervous strain, and high pressure, and that it would only get worse, his view was similar to that of Franz Kafka, the Jew who created an unfinished literary world of bureaucratic arrogance where the little man was lost and pursued, harassed and cajoled (Marshall 1933). Recent biographies have pointed out that Kafka was, in his professional life, an efficient and apparently happy bureaucrat. He could play the game.

The Bob learned American games early.[23] Like many urban boys he was fanatic about baseball statistics. Ratings and statistics; that's how we Americans tell the real from the phony. At the age of sixteen, The Bob was rating mountains of the Adirondacks for "niceness of view and all around pleasure in view and climb" (Glover 1986, 31). In college he rated convocation speakers, and in the Forest Service he kept records of pancake consumption by individual; he classified the words used by lumberjacks, noted in a statistical study undertaken firsthand, the "Manual Support to the Heads of Supreme Court Justices in Action" (Glover 1986). Bob Marshall was the master of silly statistics; he loved to play with numerology.

This was a man who chose to spend one more night out after an extended Alaskan expedition, so he would have been out for an even 50 nights; this is the man who sent printed announcements for his hundredth thirty-mile hike (Glover 1986). This was a man who knew how arbitrary the accomplishments of a man's life might be.[24]

Consequently, Bob Marshall accommodated himself gracefully to the United States Forest Service and the New Deal. With his natural facility for rating and classifying, he was expert in American games, yet always remembered that they were games.

One might read the celebrated "U-Regulations" of the Forest Service in that light. The "U-Regulations" were uniquely Marshall's, though Marshall credited Aldo Leopold with the idea of wilderness (Glover 1986). In fact, by mid-1929, Marshall had repudiated capitalist forestry as an "outmoded model" (Glover 1986, 102–103).[25] [Indeed, his biographer juxtaposes Marshall's sudden socialization with his adamant advocacy of socialism: these in turn came after a rather unpleasant anti-Semitic attack on himself and his brother, on his father and his family's honor (Glover 1986).] Yet the same Bob whose personal views of recreation were "fairly dripping with nineteenth century romanticism" could, in 1932, classify recreational areas into seven (arbitrary and perhaps ambiguous) types—Superlative, Primeval, Wilderness, Roadside, Campsite, Residence, and Outing Areas—and a year later write a book on public ownership of forests for which he considered the title, "Those Bastards, the Lumbermen" (Glover 1986).[26] The Bob's book, finally entitled *The People's Forest* was welcomed by the *Journal of Forestry* as being "a dangerous book" by a utopian radical (Glover 1986). When Marshall's "U-Regulations" were approved by Secretary of Agriculture Henry A. Wallace, nobody thought they were only some kind of silly bureaucratic game, or so we are told (Glover 1986). We hear instead of The Bob's "Immense Legacy" (Shands 1987).[27]

Silly bureaucratic game it may have been in practice, but behind the game was an unescapable reality. As early as 1928, The Bob had made the argument for "Wilderness as a Minority Right," by which he meant that an educated and cultivated minority "whatever its numerical strength, is entitled to enjoy the life it craves" (Marshall 1927,

1928). This, as he pointed out, was a Jeffersonian ideal, and many of America's most productive individuals had taken inspiration from wilderness, which was the only place modern man could find "life, liberty, and the pursuit of happiness." A cultural elitist, Marshall would make this argument in one-liners:

> "How much wilderness do we really need?"
> "How many Brahms symphonies do we need?" (Glover 1986, 116) [28]

These kinds of one-liners are typical of Marshall's style of talking, and sometimes writing. The acid tongue of the gadfly which hurts people. Good Americans do not talk like that, Marshall was told, repeatedly (see Glover 1986; Shands 1987).[29] Combined with his cultural elitism, Marshall's style branded him, as well it should. Referring to a director of the Park Service, he once wrote, "Cam [Cammerer] was his usual friendly, stupid self" (in Glover 1986, 88).[30]

However, it was clear that The Bob not only viewed wilderness as a minority right but insisted that minorities had rights to wilderness. He wanted to subsidize the transportation of low-income citizens to public forests, he wanted to create recreational opportunities for minorities, and he wanted to open federal agencies to employment for representatives of these same groups (Glover 1986).

In other words, Marshall was not much impressed by arguments about unprepared or incompetent wilderness devotees "loving the wilderness to death." [31] Nor did he like a closed technical elite in the bureaus. How could he, given his own particular ways of experiencing wild country, or his own transtechnical point of view? He was thoroughly taken with the idea of "blue-collar wilderness," but failed to convince his colleagues in The Wilderness Society.

Finally, however, one comes to an assessment of The Bob's short but energetic career. As James M. Glover observes, historians have often asserted that Marshall "single handedly" brought about a huge expansion of the nation's wilderness system, and have thus exaggerated his role (Glover 1986).[32] Marshall would probably not have appreciated being portrayed in such a role, which was not particularly appropriate. Indeed, it is precisely the role he never played during his entire career as wilderness adventurer and advocate.

In order to explain, one must use another Jewish story, which elucidates Jewish role models. We Jews remember Henry, kneeling with Nixon.[33] So it is that Joseph Heller (1979) has Gold's father say of Henry Kissinger:

> He ain't no Jew! . . . No, siree. He said he was a cowboy, didn't he? A lonesome cowboy riding into town to get the bad guys, didn't he? All by himself. Well, no cowboy was ever a Jew." (32)

And vice versa.[34]

Notes

1. The film *From Here to Eternity* (1954).

2. I was reminded of Bernard Malamud by one of my favorite writers, Wallace Stegner, in a collection of lectures, *The American West As Living Space* (Ann Arbor: University of Michigan Press, 1987), where Stegner refers to Levi Strauss as "a peddler" (76).

Although such a reference is not anti-Semitic, it is not very different from the American patrician attitude which Bob Marshall's father encountered in the late nineteenth and early twentieth centuries, the view that Jews rarely participate in basic production, that the Jew takes to the pack as the Italian to the pick (see Rosenstock 1965, 81).

Stegner believes, apparently, that Leslie Fiedler or Bernard Malamud might have been converted to a western viewpoint, if only they had been exposed to the writing of Norman Maclean, or Ivan Doig (pp. 84, 85). In an interesting turn of events, shortly after reading Mr. Stegner's lectures, I listened to Ms. Mallory Young of Tarleton State College tell the Western Literature Association that Doig and Maclean represented the slow-talking western style of which we all ought to be proud ("Writing in the Slow Lane," given at the WLA Annual Conference, Eugene, Oregon, October 5–8, 1988).

Stegner says, "It would do no harm if an occasional Leslie Fiedler came through to stir up [the West's] provincialism and set it to some self-questioning" (85). Presumably it would be best if this "outsider" didn't stay too long. Certainly he could never be one of Stegner's "stickers," who need to "get on with the business of adaptation" (86).

I consider myself one of the "stickers," and I resent implied cultural bias when I find it. Therefore, I provide my own personal testimony, in these notes, of course, read verso.

I have a friend in Cedar City, a man who owns the livestock auction. This man has a way of surprising me. Once, over dinner, he asked me how much I was worth. I told him that I had heard that the chemicals in my body were worth about ninety-seven cents. Then he told me that the average farmer in America was "worth" over $400,000. What he meant was that the farmer was in debt for that much money's worth of machinery.

Another time, at a Christmas party, I was surprised to discover that he could tell anti-Semitic jokes in my presence without being embarrassed.

As far as I am concerned, that man is a local and more provincial version of Wallace Stegner. He is my neighbor. I accept that.

3. I would prefer that "The Bob" said these words, but history does repeat itself. To give this idea Yiddish syntax:

You think these things never happened before?

They always happened before.

I will have to do some of the talking. In this "what if" argument, I play Stephen Daedalus to The Bob's Leopold Bloom. Metempsychosis. Rocks.

However, to paraphrase a character in one of the many Jewish fictions to which I refer in this text, "My name is Cohen, not Koan."

Let us imagine that the Jewish consciousness has a contribution to make in this cultural battleground we call the West. This will require a revision of the conception of land ownership. Jewish people never owned land—they were not allowed. They gravitated toward the "professions" which you can take with you into exile. The wandering Jew was the first backpacker.

Further, the central tenet of Bob Marshall's much maligned book, *The People's Forests* (1933), is that private ownership of forests has not worked, cannot be regulated, and must be ended. Our sin has been that of ownership. Like the human in the Wilderness Act's definition of wilderness, we might think of ourselves as visitors, wandering Jews (and we might, like Muir, substitute Nature for Christ in the theological myth: We are waiting, like The Bob, for the second coming of Nature.)

4. Much of what I have to say about Bob Marshall here is based on James Glover's fine research; yet my reading of Glover's book does not correspond to the meaning of Marshall's life, as Glover interprets it. Perhaps every reader puts a different interpretation on events, as I do upon Glover's book, and Glover's predecessors' understanding of Bob Marshall. I call my reading "The Bob," and I realize that the majority of readers or listeners may find my ideas full of fast talk.

5. Though Marshall was, at root, his father's son, he learned his wilderness skills from a kind of second father named Herb Clark (Glover 1986). Two fathers: one wild, the other civilized. A young urban Jewish boy doesn't learn wilderness skills from his immigrant or second generation father.

Neither could Bob Marshall learn to act like a Protestant, in 1925, when he was a summer student assistant to Richard McArdle.

As Bob Marshall's father said, in 1927, "It is the coward who seeks to be something different than what he really is, the man who is disloyal to his past and a traitor to his ancestral faith who will be despised for what he is" (Rosenstock 1965, 33). The Bob was one of those indifferent and secularized Jews whom his father disparaged, who read *Ulysses* instead of the Bible.

I remember smuggling a rock-climbing rope into my parents' house. Jews don't do those sorts of things. None of those western skills are appropriate. I learned to climb rocks from a one-legged Austrian who came to this country after the war. Bill Kittredge has the sons of the men of the West— the last of the rednecks—"climbing frozen waterfalls with pretty colored ropes," but these were not the sons of Jewish constitutional lawyers, or the sons of Austrians. Joseph Heller (1979, 32) has an eastern American father saying, "No cowboy was ever a Jew." Where does a Jew come to western wilderness without his father to guide him?

I am, in fact, a third generation American, just as Bob Marshall was. My grandfather was a furrier, as The Bob's was. My name was changed, just as the Marshall name was. My parents came from the East Coast.

Note, however, that the Marshalls were German Jews, and my people were eastern Europeans (see Howe 1976). There is as great a difference between the Marshalls and my family as between David Douglas and John Muir.

6. There is a subtle difference between The Bob and his father, Louis. On a certain level, Louis was a *Hofjude*, a paternalist perhaps, who was a Republican, because that was real Americanism and essential to assimilation of Jews into American Life; yet Louis thought that Theodore Roosevelt was a dangerous demagogue (Rosenstock 1965). On a certain level, Louis *was* a court Jew. He was an upstart, for that is what assimilation means.

And on a certain level, his son, The Bob, preferred the status of what Hannah Arendt (1978, 66) calls the "conscious pariah":

> All vaunted Jewish qualities—the 'Jewish Heart,' humanity, humor, disinterested intelligence—are pariah qualities. All Jewish shortcomings —tactlessness, political stupidity, inferiority complexes, and money-grubbing—are characteristics of upstarts. There have always been Jews who did not think it worthwhile to change their humane attitude and their natural insight into reality for the narrowness of caste spirit or the essential unreality of financial transactions.

Unfortunately, most Jews, including The Bob and his father, comprehended both the qualities of the upstart and the pariah. I shall say The Bob was perennially Green, playing on the term "greenhorn" used by Jews of my grandparents' generation.

7. "Whose god?" one might ask. We Jews can learn to cuss from Gentiles, but what does it mean when we utter the sacred names in vain?

8. Jewish mountain-climbers have always been famous for overstatement. There is Maurice Herzog, a prototype for Saul Bellow's (1964) Moses Herzog. In *Annapurna* Herzog wrote, "In my worst moments of anguish I seemed to discover the deep significance of existence which till then I had been unaware. *I saw that it was better to be true than to be strong*" (1952, 12 [italics added]). The book *Annapurna* is about wilderness as a communal resource. The climb of Annapurna is narrated not as a war (see Hunt 1954), not as a Byronic victory (see Buhl 1956), but as something else. What is this mixture of feelings? Bellow's Moses Herzog says, "I fall upon those same thorns with gratification in pain, or suffering in joy—who knows what the mixture is!" (1964, 207).

9. Fighting? Fist fighting?

10. Social embarrassment is a staple in the American Jew's life.

11. Another of Heller's characters says, "Where does a Jew come to a horse?" (1979, 95;96).

12. For literary theory see Meeker (1974). We can invent ourselves according to two models, Meeker suggests: "The tragic man takes his conflict seriously, and feels compelled to affirm his mastery and his greatness in the face of his own destruction" (22). The comic man "is durable even though he may be weak, stupid, and undignified" (23). The lessons of comedy are humility and endurance (39). Comic man knows nobody cares about him. As picaresque hero, he is an outlaw, he is an opportunist, and he is "pushy." He perceives the world as dangerous.

Yossarian, in Joseph Heller's *Catch 22* is such a hero, and so—I am arguing—is Marshall. Note that Yossarian, like Henderson the Rain King, of Bellow's novel, is a Jew in disguise. So is the author of *The Pathless Way.* Apparently there is a tradition in Jewish American letters. We deny what we are for as long as we can—that is to say until we are mature enough to stop taking ourselves seriously and admit that we are all greenhorns.

13. My heroes have never been cowboys. We're talking here about Jews in the wilderness. If you think this is a joke, it's a joke you should take seriously. Look at the picture of Maurice Herzog on the jacket of *Annapurna.* In pain, with his frozen fingers before his eyes like useless sticks, this man would say, "There are other Annapurnas in the lives of men."

Like Patty Limerick's version of Narcissa Whitman, people like Bob Marshall (or me) are *invaders,* immigrants who followed the Jeffersonian dream into the Lewis and Clark country. We might not be woodcraft rangers or cowboys, but we're smarter than some give us credit for being. And now, as third generation Americans, as postmodern Jews, we are stickers.

14. So, from the previous notes you know that first you get The Bob and then down the line you get people like me; that's a problem out West. We might not have been here first, but we keep coming back and we don't much scare, and we ruin the neighborhood, because we're not polite. America needs Eastern Jews who read fat books to say, the West isn't poor, it's cheap, it isn't strong, it's just crude, it isn't innocent, it's. . . . Well, enough. It's time to talk about Bob Marshall's ideas, the books about him, and those books he kept reading.

15. Come on, Rod. The strenuous life of Bob Marshall. A Jewish Teddy Roosevelt? Talk about counting cultural knishes as if they were as American as apple pie!

16. Fox, at least, sees Marshall as a postmodernist. He recognizes the qualities which The Bob inherited from his father, who had, according to the one biographer "a profound disbelief in the principle of collective reasoning" (Rosenstock 1965, 63).

As The Bob wrote, "Experience has always shown that if the balance hinges between satisfying material wants and aesthetic wants the choice of the people is on the side of the former" (Marshall 1933, 124).

17. What I want, of course, is a Bob I can trust. I want him as a strand of the wilderness tradition that means something and makes my life more meaningful. I want to think of his contributions as a wilderness anti-hero and as an anticonqueror of the West. I want a coherent Bob, whose real-life experience led to his view of wilderness preservation as a "minority right" and left to his insistence that access to the wilderness should be a right for minorities. I want a Bob whose impoliteness was an essential quality. I want to remember Bob as the walking bureaucrat—a "holy goof" in the age of Steve Mather and his shining touring car. I want a Bob I can laugh with, even as I laugh with Allen Ginsberg when he says "America, I'm putting my queer shoulder to the wheel."

18. Consider Moses Herzog "overcome by the need to explain, to have it out, to justify, to put in perspective, to clarify, to make amends" (Bellow 1964, 2). "He was reasoning, arguing, he was suffering, he had thought of a brilliant alternative." (*Ibid.*).

Perhaps this is not a politically useful style. As a lawyer Simkin tells Herzog, "Get a clean-cut gentile lawyer from one of the big firms. Don't have a lot of Jews yelling in the court. Give your case dignity" (Bellow 1964, 213).

19. We ought to get sloppy and emulate Bob Marshall rather than the fastidious side of John Muir, or the priggish Thoreau of "Higher Laws." We are too well-dressed in our "Patagonia" clothes. Does anyone remember what Samivel's Patagonian Climber used to look like? (Yvon does.) We are over-prepared (I'm talking contra-boyscout). We shave, we look too good, we are vain, and we never get lost. All my favorites have been slobs. Norman Clyde, for instance. Camping is not a higher form of housecleaning, or a chance for greater control.

Our eccentricities are beautiful. We are all a bit mad. So this is my confession. When I open my mouth, nobody believes that I am a Westerner, even though I was born in Dallas, Texas, raised in Los Angeles, and have lived in southern Utah for fifteen years. It must be the way I talk.

20. Who has not admired the weather which has prematurely ended a wilderness trip?

21. We are bad losers in our little tiffs with nature. When I was young and a rock climber, and I failed on a climb, which was frequent, instead of realizing that I was learning how to retreat gracefully, I felt called upon to return to camp and speak of a vendetta, as we all did, and go right back to erase my humiliation.

Well, I myself have been deracinated.

22. This is extremely difficult for a mainline professional conservationist to understand. See the review of Glover's biography by William E. Shands (1987) of the Conservation Foundation, wherein the review asserts: "Shy and slow to make friends, with a fear of women that bordered on the phobic, Marshall matured into both a ladies' man and a charismatic leader." Charisma. God's grace. Not only was The Bob not graced by a god in which he did not believe, he would have found the use of gender in the analogy offensive.

The difficulty in fitting Marshall into the charismatic tradition is underlined by the marginal comments of David Brower on a draft version of The Bob written for my *The History of the Sierra Club*. Brower disagreed when the history, following Fox, characterized The Bob as an "unabashed elitist," and Brower said of my references to Marshall's socialist tendency, "I don't believe it." As to Marshall's wrestling with modernism, the charismatic Archdruid asserted that it didn't serve the book (David Brower to Jim Cohee, February 19, 1987. In author's possession.)

23. Jews have learned American games very well indeed. It is a blind spot of a gentile like Stegner that he would imagine that the author of *The Natural* would fail to understand that "fly fishing is not simply an art but a religion, a code of conduct and language, a way of telling the real from the phony" (Stegner 1987, 84).

24. The world *we humans have created* is arbitrary at root, as any good postmodernist knows. Why not see how many mountains you can climb in a day, or make one camp more so that you can say you have been out for fifty days? Why not get out of the car and walk another three miles to make a thirty-mile day? There is no cosmic meaning in these acts, any more than there is meaning in the acts of a bureaucracy.

25. The Bob is the dialectical opposite of the magpie-killing grandfather that Bill Kittredge brilliantly describes in *Owning It All* (1987). The Bob was the man who had everything and willingly gave it away, a man who could not countenance private ownership of forests.

26. I do not consider The Bob's title inconsistent. But then I am the person who published a book in which Steven Mather is referred to as one of the "flat, tasteless bureaucrats" and as a "vulgar intellect" (Cohen 1984, 283, 310).

Mather, of course, had no sense of the absurd, and one might juxtapose The Bob, as a "walking bureaucrat" with the man who loved to tour the parks in cars; one might also juxtapose the Mather who took himself so seriously but who suffered periodic nervous breakdowns with the Marshall who enjoyed playing the antibureaucrat, writing absurd memos, and who once offered to walk to the Klamath River Reservation if the BIA wouldn't pay airfare (see Glover 1986, 162, 165).

27. Right here, right now, in Utah, we are busy playing the classification game, and not very well. There are people who line up and label themselves according to classifications, e.g., wilderness people, multiple-use people. Anyone who has read the seven volume *Utah Statewide BLM Wilderness Environmental Impact Statement* of 1986 knows how contrived in reasoning and full of silly statistics the bureau's arguments are.

Which does not mean that one does not play the game. In October 1935, The Bob did some walking in southeastern Utah, where, as he wrote, "9 million acres still remain roadless, the largest wilderness remaining in the U.S" (Glover 1986, 189). As we know, most of this land is up for grabs, and there are plenty of people who want to grab it. Some of my neighbors, for instance.

28. Leo Rosten (1968) shows that the use of a question to answer a question is a typical American colloquial borrowing of a Yiddish linguistic device. It is a satirical strategy used to indicate that the answer is self-evident and constitutes an affront.

29. Indeed, Marshall talked and wrote just like characters in Bellow's Jewish fiction, or in Heller's *Good as Gold*, where the typical strategy in an argument is to go for the jugular. This *is* a cultural trait.

30. This author would never say such a thing.

31. For "loving the wilderness to death," see Nash (1967, 316–341). Nash's idea is probably an ethnocentric notion, but then probably so is wilderness, as we have come to legislate it.

The Nash argument comes from Olaus Murie, who—in 1940—feared that if too many people were brought to the wilderness "there would be an insistent and effective demand for more and more facilities, and we would find ourselves losing our wilderness" (in Glover 1986, 219). The use of pronouns here is revealing, but Nash turns Murie's reasoning into slogans and somehow makes it offensive, as it was not in the original. Shouldn't a historian be more perspicacious than his subject?

In a strange turn, Murie's argument becomes Nash's and then one finds Richard McArdle using it (see McArdle 1975).

32. See also Wild (1979), Shands (1987), and others. The simple fact, as Glover (1986) notes, is that Bob Marshall has been spoken of in a language he found repulsive.

33. We remember the Henry Kissinger who said, "Power is the great aphrodisiac." We remember the one who compared himself to "the cowboy leading the caravan alone astride his horse, the cowboy entering a village or city alone on his horse" (see Heller 1979, 302–303).

This matter of allegiances, a problem compounded by the rise of Zionism in the twentieth century: we do not wish to kneel with Nixon, but in a situation where American Jews are accused of having allegiance to no land—or having dual allegiances—there is tremendous pressure to live a lie by acting as if one were a "real" American.

Irving Howe (1976, 646) speaks of Eastern European Immigrant Jews:

Their greatest contribution has been less what America has accepted than what it has resisted: such distinctive traits of the modern Jewish spirit as an ever eager restlessness, a moral anxiety, an openness to novelty, a hunger for dialectic, a refusal of contentment, an ironic criticism of all fixed opinions.

34. And now for the *caveat*.

As James M. Glover (1986, 131) points out, Robert Marshall saw himself as "prejudiced in favor" of Indians and did not understand the implications of such a view. Further, believing that he could solve problems *for* Indians, Marshall, as forester for the BIA, prescribed a "wilderness solution to Indian woes" (see Limerick 1964; and Collier 1964). In other words, in his dealings with Indians, Marshall was ethnocentric and paternalistic; he may not have played the cowboy, but he was his father's son.

The Bob I have depicted here is not a hero but an antihero; the example of his life offers a corrective, not a final solution to certain problems in the conservation community. It is even possible that The Bob drove himself to death. As many said, and this too applies to the author, "He never rested."

References

Arendt, H. The Jew as pariah: Jewish identity and politics in the modern age. Feldman, R. H., ed. New York: Grove Press; 1978.

Bellow, S. Herzog. New York: Viking; 1964.

Buhl, H. Lonely challenge. New York: E. P. Dutton; 1956. Translated by H. Merrick.

Cohen, M. P. The pathless way: John Muir in American wilderness. Madison: University of Wisconsin Press; 1984.

Collier, J. Wilderness and modern man. In: Brower, D., ed. Wildlands in our civilization. San Francisco: Sierra Club; 1964: 115–127.

Fox, S. John Muir and his legacy: The American conservation movement. Boston: Little, Brown and Company; 1981.

———. We want no straddlers. Wilderness (Winter):5; 1984.

Glover, J. M. A wilderness original: The life of Bob Marshall. Seattle: The Mountaineers; 1986.

Heller, J. Good as gold. New York: Simon and Schuster; 1979.

Herzog, M. Annapurna. New York: E. P. Dutton; 1952.

Howe, I. World of our fathers. New York: Harcourt Brace Jovanovitch; 1976.

Hunt, J. H. B. The conquest of Everest. New York: E. P. Dutton; 1954.

Kittredge, W. Owning it all. St. Paul, MN: Graywolf Press; 1987.

Limerick, P. Legacy of conquest. In: Brower, D., ed. Wildlands in our civilization. San Francisco: Sierra Club; 1964: 200–209.

Marshall, R. Wilderness as a minority right. United States Forest Service Bulletin (August):5–6; 1927, 1928.

———. The problem of the wilderness. Scientific Monthly (February):141–148; 1930.

———. The people's forests. New York: Harrison Smith and Robert Haas; 1933.

McArdle, R. E.; Maunder, E. R., Wilderness politics, legislation and forest service policy. Journal of Forest History 19:176–179; 1975.

Meeker, J. W. The comedy of survival: Studies in literary ecology. New York: Charles Scribner's Sons; 1974.

Meine, K. Aldo Leopold: His life and work. Madison: University of Wisconsin Press; 1988.

Nash, R. Wilderness and the American mind. New Haven, CT: Yale University Press; 1967.

Rosenstock, M. Louis Marshall, defender of Jewish rights. Detroit: Wayne State University Press; 1965.

Rosten, L. The joys of Yiddish. New York: McGraw Hill; 1968.

Shands, W. E. Review of A wilderness original: The life of Bob Marshall. Environmental Review (Summer):147–149; 1987.

Stegner, W. The American West as living space. Ann Arbor: University of Michigan Press; 1987.

Wild, P. Bob Marshall, last of the radical bureaucrats. High Country News. 1979 Mar. 9.

The Wilderness Within Us
Women and Wilderness

FELICIA F. CAMPBELL

"I HAVE homesickness for a country that isn't mine. The steppes, the solitudes, the eternal snows and the big skies haunt me," wrote famed explorer Alexandra David-Neel (Foster and Foster 1987, 128–129), mourning her separation from the Himalayan wilderness. "A part of you will always keep a bit of our world," Milagros tells Florinda Donner as she leaves the *shobono*, a village deep in the jungle between Venezuela and Brazil (Donner 1983, 301). "You will never be free."

Fitting easily with the native populations, reveling in freedom from the restrictions and demands of traditional society, glorying in veldt or mountain peak, such women as these found their true homes in wildernesses far from the Western world in which they were born. For such women, no matter how great their material success or how much adulation they receive from the public, their wilderness experiences proved the most successful and rewarding of their lives. For them, the creature comforts and trappings of the so-called civilized world to which they returned provided little more than upholstery for the prison that is "civilization."

In the comfort of such a prison, I edit and reedit this article. From the window at my back, Nevada's Red Rock Cliffs reflect in the screen of my computer, nature superimposed on technology. I do not wish to give up either. The Red Rocks serve as stand-ins for my beloved Karakorams. How does one express the inexpressible?

"I taste a liquor never brewed," wrote poet Emily Dickinson who, by an effort of will created wilderness in her own back yard, seldom venturing beyond her garden gate. Alexandra David-Neel, whose earliest desires were to see what lay beyond the garden gate (1927), would have understood.

This student activist turned diva, then scholar, Tibetan lama, and explorer, who lived from 1868–1969, dwelt on the sensual, seductive quality of her romance with the wilderness. In a marriage that existed largely by correspondence, she wrote her husband, "Sensuality is to each in his own way. . . . To me, it is being alone, silence, virgin land not disfigured by cultivation, vast spaces, a rude life under a net" (in Foster and Foster 1987, 170).

A year in a Himalayan hermitage had confirmed this emotion. "Solitude! Solitude!," she wrote (David-Neel, 1965, 78). "Mind and senses develop their sensibilities in this contemplative life made up of continual observations and reflections. Does one become a visionary or, rather, is it not that one has been blind until then?" She knew that she could not remain in her hermitage forever, but suffered an indescribable terror when she thought of "the sorrowful world that existed beyond the distant hill ranges" (79).

Hunger, cold, and physical suffering she accepted as necessary evils. In a letter (Foster and Foster 1987), she wrote: "These miseries pass quickly (and) one remains permanently engulfed in the silence where only the wind sings, in the solitudes almost naked of greenery, the chaos of fantastic rocks, dizzying peaks and horizons of blinding light" (155).

From her first look at the Himalayas (David-Neel 1927), she knew that she had found her place. "What an unforgettable vision! I was at last in the calm solitudes of which I had dreamed since infancy. I felt as if I had come home after a tiring cheerless pilgrimage" (xi). In fact, she became so at home in Tibet that at age 54, disguised as a beggar woman and accompanied by her adopted son Yongden, she had walked the four-month journey from the China border, through villages and over mountain passes, to the forbidden city of Lhasa. This fifth journey led to the publication of *My Journey to Lhasa* in 1927 which increased her already considerable fame.

The Himalayas were to remain David-Neel's home even after she returned to France at age 78, settling in her home at Digne which she named Samten Dzong, Fortress of Meditation. The Fosters, her biographers, tell us (1965) that she called the surrounding French Alps "Mountains for Lilliputians" (262). Here was no awe inspiring spectacle such as that she described in *My Journey to Lhasa* (1927) which would "make believers bend their knees, as before the veil that hides the supreme face" (128).

The Fosters (1965) believe that she had accepted the path of the bodhisattva, putting aside her own desires to share what she had learned. It was a heavy price to pay. She regretted that she had not died in her hermitage (Foster and Foster 1965, 323). "I arrived there at the summit of my dream, alone, soaring like an eagle, in my cavern on a Himalayan peak. What remains for me to do, to see, to experience—After that?" Had she died there, the old lama under whom she was studying would have carried her body still higher and left it there. As she said, it would have been "simple, not banal," and according to her desires.

Julie Tullis was to die on her Himalayan peak, her mountain of mountains, K2. Her very personal relationship with that mountain brought her knowledge, fulfillment, and ultimately death after she had achieved its summit. In her autobiography, *Clouds from Both Sides*, written in 1986, the year of her death, and published in 1987, she describes K2, which had taught her "the real pleasures of high altitude mountaineering" and "totally captivated her" (1987, 5). Explaining that even had she reached the summit on the first try, she would have wanted to return, "to discover more about it, explore its other sides." It was the one "most special" to her.

Tullis likened her passion for mountains to that of a sailor for the sea. She wanted to be with them, exploring each for its character and moods (1987). Mountaineering was for her "a way to achieve a harmony with nature" (1987, 220). She was drawn back to K2, her "mountain of mountains" when she knew that each attempt shortened her chances of survival, because "life is short and there has to be a reason to live beyond purely surviving, an extension of that force of nature of which my body and spirit are a part, just as a river has to flow and broaden or it will stagnate" (1987, 22).

Unlike Alexandra David-Neel, Julie Tullis died in the spot she would have wished. A few months before her death, she wrote, "If I could choose a place to die, it would be the mountains," explaining that she had learned while falling in an avalanche that she would not mind dying like that. There were other occasions "when just to sit still and drift in an eternal sleep would have been an easy and pleasant thing to do" (1987, 276). Her wish that nature's circle would not close for her too soon was not granted.

Florinda Donner, a United States citizen born in Venezuela of German parents, went not to the heights of the Himalayas, but to a *shobono*, a village deep in the jungle between Venezuela and Brazil. Like David-Neel, Donner is a scholar, an anthropologist whose major interest is shamanic healing.

In *Shabono* (Donner 1982), she describes a year spent among the Iticoteri, a tribe of Yanomamo Indians. During that year, Donner, a small blond woman, was adopted by an Iticoteri family and became totally immersed in their culture. Her youth, her slightness of stature, her being a woman alone meant that she posed no threat to them. In addition, the Iticoteri believe "that whites are infantile and thus less intelligent" (Donner 1982, xi), and so cared for her, attempting to educate her and initiate her into their ways.

There she learned "the secrets of the forest—secrets of misty caves, of the sound of sap running through the branches and trunks, of spiders spinning their silvery webs" (Donner 1982, 299). These secrets create in their possessor an eternal nostalgia for the jungle. The yearning would not be hers alone. The Iticoteri would remember her as well. Iramamowe, a shaman, told her in farewell, "There will be nights when the wind will bring your voice mingled with the cries of monkeys and jaguars. And I will see your shadow on the ground, painted by moonlight. On those nights, I will think of you" (1982, 292).

In the end, Milagros, who had taken her into the jungle, reappears at the mission where she is awaiting transport to take her out of the jungle and back to Los Angeles. She tells him that her experience with his people was "like a dream." He replies, "A dream that you will always dream. A dream of walking, of laughter, of sadness. . . . Even though your body has lost our smell, a part of you will always keep a bit of our world. . . . You will never be free" (1987, 301).

Karen Blixen (Isak Dinesen, 1855–1962) is another who was fated never to be free. She found her home in the African wilderness and alienation outside it. The yearning is evident as she opens *Out of Africa* (1985, 3). "I had a farm in Africa." While her love affair with Denys

Finch-Hatton is most famous, largely because of the recent movie, her love affair with the land was perhaps even greater. In Africa was a lightness of being; one could breathe. She describes the air as a chief feature of the landscape.

> Looking back at a sojourn in the African highlands, you are struck by the feeling of having lived for a time up in the air . . . you breathed easily . . . drawing in a lightness of heart. . . . you woke up in the morning and thought: here I am where I was meant to be. (1985, 4)

Of a safari with Denys Finch-Hatton, she wrote:

> I knew then without reflecting, that I was up a great height upon the roof of the world, a small figure in the tremendous retort of earth and air, yet one with it; I did not then know that I was at the height of my own life. (Dinesen 1985, 443)

Her return to Denmark after she had lost the farm marked the beginning of an endless exile. Like David-Neel, she was homesick for a country that wasn't hers. According to her biographer Judith Thurman (1982), she found it almost impossible, except with her animals and servants, to achieve a responsible and steady relationship.

Speaking before the American Academy of Arts and Letters as an honorary member, she divided her life into five segments. The third, that in Africa, she called "her real life." Essentially, in Africa, she was not lonely. "My daily life out there was filled with answering voices. I never spoke without getting a response; I spoke freely and without restraint, even when I was silent" (in Thurman 1982, 466).

She did not take on the role of the bodhisattva as did David-Neel, yet her writings based in Africa have moved millions.

Perhaps, I think, it is not given for us to remain in the wildernesses. Perhaps the essence of the experience is that it is taken from us, yet provides a touchstone for all that we do from then on in that everything that happens thereafter is measured against those moments of unity. Perhaps, too, if we remained, those wildernesses would cease to be.

Beryl Markham does not quite fit the template that I am using, but she nonetheless deserves to be included here. Living in Africa from age four made her almost a native, not a visitor from another realm. Her upbringing, running wild on her father's farm, playing with the native children, learning to hunt and carry a spear gave her a perspective that would forever place her outside the colonial culture in which she moved with such elegance.

Undoubtedly Markham loved Africa. "It is mystic; it is wild; it is a sweltering inferno; it is a photographer's paradise; a hunter's Valhalla; an escapist's Utopia" (1983, 8). She explains that, defying interpretation, it is all things to all people. For many like herself, it is home. "It is all these things but one—it is never dull." Never able to discuss boredom until she had lived in England for a year as an adult, she decided that "boredom like hookworm is endemic." David-Neel, Tullis, Donner, and Blixen would agree.

Always returning to Africa, Markham died there in her eighties having trained winning race horses until a few years before her death. The first female professional pilot in Africa, her autobiographical *West with the Night* so captured public attention on its reissue in 1983 that Markham became the subject of both a Public Television Documentary and a bad TV miniseries that spent more time on her famous love affairs than on her truly remarkable career as a pilot. She was the first person to fly solo from Britain to the United States.

It is Markham's passion for the air that is important here. David-Neel, Dinesen, and Tullis all experienced lightness and exaltation in the space around them, but Beryl Markham was gifted with flight. "The air takes me into its realm," she exalted. "Night envelopes me entirely, leaving me out of touch with the earth, leaving me within this small world of my own, living in space with the stars" (1943, 15).

Her Gypsy Moth airplane was an object of love, perhaps her greatest object of love. "I feel through the soles of my feet on the rudder bar the willing strength and flex of her muscles. . . . The wind in the wires is like the tearing of soft silk under the blended drone of engine and propeller. Time and distance slip smoothly past the tips of my wings" (1943, 16).

So here we are, my subjects and I, time and distance slipping smoothly past. The wonder of the Himalayan peaks and the joy of dancing with the Hunzakut under the stars in Karimabad in a circle below the Mir's palace remain with me, as does the reflection of Red Rock on the screen before me.

I miss the warm comradeship of my guides Sarbaz and Abbas on our mutual enterprise to march my out–of–shape, middle-aged body

through the 150 miles of the Karakoram range of the Himalayas to K2—
my adventure, their livelihood. I miss the sustaining hands of the porters
when I needed help and their applause when I didn't, yet they are never
truly absent. The distant silences and solitudes sustain me in the midst
of city traffic.

Still, I too am homesick for a land that isn't mine. Like the others, I
write. First the memoir, then the tales. I can't stop. Writing fills the gap.
Like the others—like Alexandra David-Neel who renewed her passport
at age 100—I plan another journey. Maybe next year.

References

David-Neel, A. My journey to Lhasa. New York and London: Harper and
Brothers; 1927.

———. Magic and mystery in Tibet. New Hyde Park, NY: University
Books; 1965.

Dinesen, I. Out of Africa [1937] *and* Shadows on the Grass [1961]. New
York: Vintage; 1985.

Donner, F. Shabono. London: The Bodley Head; 1983.

Foster, B.; Foster, M. Forbidden journey: The life of Alexandra David-Neel.
New York: Harper and Row; 1987.

Lovell, M. S. Straight on till morning: The biography of Beryl Markham.
New York: St. Martin's Press; 1987.

Markham, B. West with the night. Berkeley, CA: North Point Press; [1943]
1983.

Thurman, J. Isak Dinesen: The life of a storyteller. New York: St. Martin's
Press; 1982.

Tullis, J. Clouds from both sides. Great Britain: Gafton Books; 1987.

Toward a Holistic Eco-vision
The Infusion of the Eco-feminine in Eco-philosophy

NEILA C. SESHACHARI

M O S T scholars agree that the world today is in peril. There is a general feeling that we are on the edge of a global catastrophe and that it has been brought about by our indiscriminate advances in technology and in the natural and physical sciences. The dramatic and brutal ushering in of the atomic age by the United States in August 1945 led to a crisis of confidence not only in the United States itself but also in human nature. As John O'Neill (1985) points out, ours may be the first civilization that thinks of itself as the last. Even without war and willful destruction of the human population, our so-called technological advances have begun the task of destroying the earth. Our nuclear wastes have reached dangerous levels of safety for Earth's inhabitants; Earth's protective ozone layer has eroded perilously and even developed a hole. Plant and animal species are becoming extinct at an alarming rate; our continuing pollution of the environment has pushed living conditions to dangerous levels. Our social and political lives are becoming more strife-ridden as well.

Emergence of Ecological Concerns

In a world where countries have become increasingly competitive and power hungry at one level and progressively interdependent at another, the global situation seems headed toward certain destruction. Even so, there has emerged in the last two decades or more a small but growing voice of concern for preservation of the earth and the life it supports.

Today we are at least talking about the dangers of exploitative and unrestricted "progress in all spheres." Ecological concerns loom large in practically every discussion of industrial development and technological/military research. From the idle talk about the dangers of environmental pollution that characterized our national scene about twenty years ago, some parties involved significantly in production—the government, giant corporations, scientists—are at least voicing a need to avert the dire consequences of our irresponsible technological and industrial growth of the last hundred years, and some are taking positive action. The circle of environmental crisis is indeed closing upon us fast and we must act speedily to avert irreversible ecological disasters that endanger the very existence of life on Earth.

At the practical level, altruism has not motivated an environmental clean-up. Rather legal suits have begun to achieve what ecologists and eco-philosophers could not. They involve billions of dollars from victims of chemical and nuclear wastes, acid rain that seriously threatens agricultural crops, and numerous related problems.

But removing pollution is not enough. Cleaning up the environment has been compared to "picking up garbage" without controlling the source of the garbage. We Americans know how difficult and economically well nigh impossible it is to clean up the environment and dispose of dangerous wastes accumulated even during the short time we have produced such wastes. In addition, we cannot genuinely rid ourselves of some radioactive pollutants except by sending them into outer space, a remedy perhaps riskier to Earth than the pollution problem itself. Put simplistically we cannot contain the pollution problem by putting cellophane tape at its bulging seams. Eventually we must redefine our material "needs" to reflect the conservation needs of our global environment. And yet, it is naive to think that we can control society's material needs or legislate scientific and technological research; we cannot really turn the clock back. Sensitizing our collective conscience remains the best recourse we have.

Task at Hand

Thus our task, it seems to me, is two-fold. 1) The first part consists of taking practical steps to clean up both the environment and the sources that create environmental pollution, so that they do not threaten any life—human, nonhuman, or plant life—on Earth. 2) The equally important other task involves realigning our thinking processes to counteract the values of domination and subordination of nature that have been ingrained in our culture. Not until it espouses new values of reverence and preservation of nature will our technological civilization begin the task of reversing suicidal tendencies fostered in its own collective psyche.

A number of environmental protection groups, philosophers, and eco-activists have already begun the stupendous task of retraining our attitudes toward nature. Proponents of deep ecology, elucidators of "perennial philosophy" geared to new ecological awareness, and eco-feminists have done much to redefine our values vis-à-vis nature. An ecological consciousness has been forced upon us and a ground swell favors environmental protection both at the level of the general public and the technocratic world. Phrases like environmental ethics, eco-philosophy, and eco-feminism attest to the resurgence of interest in ecology and the environment. At the abstract level of symbolic interaction the very meanings of these concepts are changing or being modified through a subtle process of personal and communal social interaction.[1] Our age-old beliefs about nature and the meanings we bestow upon these beliefs are undergoing a dramatic change through a dual process of interaction with others and with ourselves.

Place of the Psychological Feminine in Eco-philosophy

However, an erstwhile, surface ecological consciousness will not solve our problems in human affairs. I believe that the real malaise of our civilization lies at the root of its unnatural subordination, fear, and rejection of the psychological "feminine." Further, unless this "feminine" in its genuine fullness is elevated as a positive value and incorporated in both its collective psyche and eco-philosophy, our culture will continue to have ambivalent and schizophrenic attitudes toward human interdependence on nature—seen both in its animate and inanimate manifestations—and on nonhuman life.

Amidst predictions of gloom, however, there is hope for humanity in the growing ecological concerns voiced all around. This hope is being ushered in, not through momentous scientific discoveries as in the past, but through fundamental changes in human relationships and attitudes. These in turn are changing the very meanings that "nature" and "ecology" hold for us. Since the willful abuse of our natural environment is at the heart of our problems, the catch-all for this caring approach seems to be ECO, as evidenced in our sudden interest in eco-philosophy, eco-piety, eco-science, eco-everything! These modifications in meanings have ushered in new values and new agendas for action. The old paradigm of conquering and vanquishing the earth for "man's progress" is giving way to a new one which includes ecological preservation and "human welfare." Even our vocabulary has changed. Symbolically, this change in vocabulary signifies soul-shaking changes in the human psyche.

The work of feminists who have infused the feminine principle into the traditionally masculine or patriarchal paradigms that have come down to us in every sphere of human endeavor lie at the root of this momentous change. The feminine principle, in its larger symbolic application, is the infusion of ideas of preservation, care, relationships, and the promotion of win-win situations in place of win-lose propositions in practically every human endeavor in the natural and biological sciences, medicine, social sciences, psychology, and family life.

The infusion of the so-called feminine into the patriarchal masculine gives humanity a psychic wholeness which promotes a holistic approach to human affairs. Even though I myself am a feminist, I also hope to argue that feminism as a philosophy has not yet come to terms with the psychological feminine. Recognizing that relationships, caregiving, and love were the very feminine attributes that led to women's subordination and bondage in the past, feminism has yet to take a viable stand on the interdependence of women, men, and children for psychic well-being and wholesome living.

A reverence for nature has not been part of the religiocultural heritage of the West. Therefore, the more difficult task seems to be in realigning or retraining human thinking to incorporate values that have so far never been part of our religiocultural heritage. These values pertain to the recognition of the interdependence of all forms of life and nature, whether animate or inanimate. Contrary to values of Western civilization that historically have led to misogyny and patriarchy, a belief in interdependence is necessarily nonhierarchical and egalitarian.

Misogyny and Exploitation of Nature

Of the complex reasons for the misogynous and patriarchal nature of our culture, three have been particularly influential in shaping our intellectual and cultural values. 1) Pre-Christian Greco-Roman philosophies developed on dualistic, divisive lines. The Greek philosopher, Paracelsus of Elia (sixth century B.C., for instance, categorizes all knowledge into "The Way of Truth" and "The Way of Opinion." This dualistic mode of categorizing all things into polar opposites—like good *versus* evil or light *versus* darkness—eventually led to the duality of man *versus* woman and the elevation of man as the superior and the demotion of woman as the inferior of the human sexes. 2) Early Orthodox Christianity not only continued the tradition of recognizing male supremacy, but added another potentially dangerous belief—that the earth was created for man's benefit—which sanctioned the exploitation and plundering of nature. 3) Finally, the Protestant work ethic, which developed much later in the sixteenth century, linked material prosperity to ideas of spiritual salvation, thereby subordinating spiritual values and promoting economic expediency. We see the legacy of these three cumulative beliefs in the unbridled exploitation of nature and consumption of goods and in questionable business ethics which take shelter behind the holy cow of capitalism.

The development of science in the seventeenth and eighteenth centuries continued the patriarchal traditions of domination both over the female and over nature. It completely severed any sense of responsibility for human welfare from the Renaissance-spawned yearning for knowledge and personal glory.

Thus misogyny and exploitation of nature get inextricably bound in Western male psyche. Devaluing the female and feminine compounds the problem by inflicting deep, psychic injuries on the male self. What our patriarchal culture has designated as masculine and feminine qualities are only nomenclatures, not attributes and impulses confined to males and females respectively. The masculine and feminine form the pool of human qualities; both contribute to the development of psychic wholeness in any human being, male or female. By devaluing and cutting off the sap of the so-called feminine within his own being, man has inflicted deep psychic injuries on both himself and the female. Like a color-blind individual, man is not even able to see the value component of the psychological feminine—qualities like nurturance, preservation, care, and symbiotic relationships. For him, the feminine is an inferior realm and therefore nonexistent as a personal value.

The result of this lacuna in male psychic development has vitiated the development of science and philosophy in our civilization. Well-meaning and amicable scholars often ignore half of humanity in researching human development.[2] When philosophers posit their theories, they not only tend to ignore the female sex but unwittingly deny full stature even to the collective male consciousness by banishing half the human traits of the male sex.[3] This unnatural suppression of half of our human traits has impoverished our abilities to evaluate our own problems and define our collective vision. Only restoring the so-called feminine in the male psyche (as well as masculine in the female psyche) can bring hope of a holistic vision to our culture. This holistic vision could be temporarily described as an androgynous vision. The term androgynous, which etymologically means that which has reconciled the (opposites of) male and female, is ultimately a fallacious term, because there are no psychological traits that are exclusively male or female. However, androgynous—*andro* meaning male and *gyne* meaning female—signifies the elevation of the so-called feminine to a status equal to that of the so-called masculine, just what eco-philosophy today needs.

Loss of Values and Environmental Crisis

Our environmental crisis has been attributed to loss of values in our culture. However, not many philosophers recognize that the one value that has never been accepted into our corporate value system and which we need desperately is the feminine in its fullness.

In his insightful book titled *Eco-Philosophy: Designing New Tactics for Living*, Henryk Skolimowski (1981) posits that a separation of knowledge and values has made ours a schizophrenic civilization which has lost its faith, confidence, and direction. He deplores the fact that somewhere in the development of Western thinking, knowledge and values get hopelessly separated. In classical antiquity as exemplified by Plato, Skolimowski notes that knowledge and values were fused—one is not dominant or subservient to the other. In the Middle Ages, knowledge was still fused with values but was subordinate to them; these values, Skolimowski points out, were those sanctioned by the Church. Then in the post-Renaissance period, knowledge gets separated from values without any hierarchy between the two. Immanuel Kant (1720–1804) summarized the autonomy of both these realms succinctly when he declared, "The starry heavens above you and the moral law within" (in Skolimowski 1981). The fourth and last position, says Skolimowski, is the one held by classical empiricism and its more recent extensions: nineteenth-century positivism and twentieth-century logical empiricism. This position is characterized not only by the separation of knowledge and values, but by the supreme importance given to knowledge.

Eco-philosophical Approaches

Skolimowski presents his conception of value-enriched eco-philosophy as an antidote to the environmental abuse in our culture. However, in his excellent conceptual mandala of eco-philosophy, he nevertheless overlooks "the fullness of the psychological feminine," which stands for nurturance, caregiving, and joy in the knowledge of interdependence with significant others (including nature and nonhuman life, in this instance). Because these qualities have been devalued in our culture, and because their lack has significantly impoverished it as well, it behooves new philosophies to include them specifically to reinstate them as significant values. It would not do to argue that these values are included or implied in others. As eco-feminist scholars have pointed out, our environmental crisis has psychosexual roots, which in turn can be traced to misogyny. We would not have created the environmental crisis if our culture had valued the psychological feminine.

Another significant development for our environmentally beleaguered condition is Arne Naess' concept of ecosophy (wisdom with regard to ecology), which eventually crystallized into the eight basic principles of deep ecology as propounded jointly in 1984 by Naess and George Sessions (see Devall and Sessions 1985). These tenets are both terribly attractive and flagrantly radical in the sense that they appeal to all nature lovers but are in themselves not quite practicable. Deep ecology is deeply democratic in that it posits that all creation—human and nonhuman creation as well as nature—has intrinsic worth. Therefore, it argues, nature should not be considered as a resource for humans; by the same argument, humans should not resort to dominance over nature. Also, humans should not work for an increased standard of living but should pursue quality of life.

Thus, while giving intrinsic rights to nonhuman life and nature alongside humans, deep ecology at times appears to be downright dictatorial toward the human species. What exactly do Naess and Sessions mean when they advocate that humans should pursue quality of life? At best, quality of life is an idealistic, elusive term which has evaded many of its votaries including myself. Is there a measurable standard of living appropriate to all humans? Do humans have the right to fall below or rise above this arbitrary standard, if somebody were to devise one? Naess went on record in a 1981 interview in Los Angeles, saying, "I should think we must have no more than 100 million people if we are to have the variety of cultures we had one hundred years ago. Because we need the conservation of human cultures, just as we need the conservation of animal species" (in Devall and Sessions 1985).

If drastic cuts in the human population were possible, who amongst the teeming five-plus billion humans of this earth would be given the order to vanish? Such prescriptive principles are themselves steeped in the patriarchal tradition of our culture. Patriarchy sanctions domination and prescription. There is another way of achieving the same goals of ecological preservation and cultural conservation through workable, acceptable alternatives. These goals will not be achieved by advocating the unrealistic claims of certain political groups who think that no matter how much our population explodes, the earth is bounteous enough to produce more food. However, ecological preservation, including population control, can promote a win-win situation for both humans and their environment. These mutually beneficial strategies can be introduced in gentler, kinder ways—through the infusion of the psychological feminine in the traditional, prescriptive masculine mode of problem solving. The psychological feminine would change the prescriptive paradigm into one where care, mutual respect, preservation, and interdependence would substantively change our values and goals to bring about longer lasting results.

Deep ecology has gone out of its way to develop ecological consciousness within the contexts of different spiritual traditions—Christianity, Taoism, Buddhism, Native American rituals and the like. It "derives its essence," as Devall and Sessions (1985, 80) point out, from "the perennial philosophy, the pastoral/naturalist literary tradition, the science of ecology, the 'new physics,' some Christian sources, feminism, the philosophies of primal (or native) peoples, and some Eastern spiritual traditions." This declaration is genuinely inspirational to all who are disconcerted at the general close-mindedness of Americans. And Devall and Sessions (1985, 80) further assert, "In a certain sense it [deep ecology] can be interpreted as remembering wisdom which *men* once knew" (emphasis added). In a book that scrupulously avoids sexist language, it is difficult to conjecture whether this use of the term men is generic or genuine. In any case, it may be termed Freudian to emphasize the deep subconscious level from which it surfaced.

Our culture has never, at its deepest level of conviction, considered wisdom as emanating from anybody or anything other than man. This belief is the ultimate put-down of the female and of nature, not to mention nonhuman life. The remedy for this kind of lopsidedness in our psyche is the conscious infusion of the feminine, as well as the ecofeminine, in our philosophies—eco-philosophy, ecosophy, environmental philosophy, or perennial philosophy.

Eco-feminism and the Eco-feminine

A word about eco-feminism. Feminism has indeed been a liberating force of the highest magnitude in our century. Feminists have crusaded tenaciously to restore to women full human rights and all other rights considered the birthrights of all males. In the process, they have exposed the patriarchal bias—specifically that of regarding the male as the norm and the female as the derivative or deviant—in centuries of so-called scientific investigations, which in many cases negated the research itself. Feminists have exposed the misogyny that prevails globally in patriarchal cultures and the psychosexual roots of many of our social ills. Eco-feminism specifically has revealed how many of our environmental crises have psychosexual roots. It has helped us see how sexism (mind and body pollution) is fundamentally linked to ecological destructiveness (environmental pollution). There is no doubt that feminism has enriched the perspectives of ecology by illuminating the "masculine" nature of modern science that ultimately leads to environmental destruction.

But we need to distinguish between feminism, which is a political philosophy like capitalism or Marxism and the psychological feminine which is a universal human attribute. Historically, feminism has had problems coming to terms with the feminine. At one level, feminists have argued vigorously for the recognition of the value of the feminine—of nurturance, relationships, motherhood, and the like. At the same time, feminists themselves have had very ambiguous attitudes toward women's roles as nurturers, wives, and mothers. Recognizing that the so-called feminine qualities dumped on women willy-nilly by men were the cause of their own domination, feminism has had problems with issues dealing with both motherhood and relationships with men. In some radical feminist circles, even heterosexuality is regarded as politically incorrect! It would be intellectually difficult to equate feminism with the feminine. However, feminism has brought value to the female and the feminine in human considerations. If the female and the feminine are beginning to be put on equal footing with the male and the masculine in psychology, it is no thanks to Sigmund Freud and his ilk.

Eco-feminine As Fostering a Holistic Eco-vision

The infusion of the so-called feminine into the patriarchal masculine gives humanity a psychic wholeness which promotes a holistic vision in human affairs. Feminism is the mother of this new holistic vision, but this offspring of feminism and patriarchy, itself an entirely independent and vital force, is already changing our ways of thinking and has the potential to save the world, literally and psychically.

Notes

1. Herbert Blumer, in *Symbolic Interactionism* (1969), argues succinctly how and why only stimuli, attitudes, and motives (conscious or unconscious) are not enough to explain changes in human values and attendant (human) conduct. He points out how meanings that things hold arise out of the social interaction that one has with one's species, and that these meanings are modified through interpretative processes that individuals use in dealing with things they encounter.

2. Educational psychologist Carol Gilligan's *In a Different Voice* (1982) eloquently points out the sexist bias in educational research. Kohlberg, whose theories had become the canon in educational evaluation, for instance, conducted his longitudinal studies in human maturation levels by evaluating eighty-four boys and no girls. Basing his observations on responses given by boys, he came to the conclusion that girls do not ever achieve the level of maturity that boys do.

3. I have discussed these ideas in detail in "Toward a Gender-Based Equality in Society: Reflections on Masculinity, Femininity and Androgyny" in Seshachari (1989).

References

Blumer, H. Symbolic interactionism: Perspective and method. Englewood Cliffs, NJ: Prentice-Hall; 1969.

Devall, B.; Sessions, G. Deep ecology: Living as if nature mattered. Salt Lake City, UT: Peregrine Smith Books; 1985.

Gilligan, C. In a different voice. Cambridge, MA: Harvard University Press; 1982.

O'Neill, J. Five bodies. Ithaca, New York: Cornell University Press; 1985.

Seshachari, N. C. Toward a gender-based equality in society: Reflections on masculinity, femininity and androgyny. In: Conover, R.; Holt, R. L., eds. Great debates and ethical issues. Ogden, UT: Weber State College; 1989: 39–60.

Skolimowski, H. Eco-philosophy: Designing new tactics for living. Great Britain: Marion Boyars Publishers Ltd.; 1981.

Wilderness References in Urban Landscapes

ANNE BERMAN

EDWARD B. WEIL

T H E R E is a tradition in the architecture of various ancient cultures to site monuments, structures, or a collection of structures in pristine wilderness locations. Greek temples set in the craggy terrain of the Mediterranean and Incan cities perched in the Andes are familiar examples. Built form in these circumstances serves as a paradigm for human order and reason, a concept accentuated by its juxtaposition with a rugged untamed landscape. The wilderness serves as a spiritual link to divinity and the forces of nature which are beyond the control of mankind (Kostof 1985).

The relationship between the city and the wilderness has been explored in historical (Nash 1982), philosophical (Weinstein 1982), and socio-spatial (Tuan 1971; 1974) treatises. While perhaps seeming incongruous, ultimately the perception of one depends upon perceptions of the other, as well as upon perceptions of transitional or rural landscapes. A brief restatement of Nash's detailed history provides temporal contexts for the American ideological transformation. Nash's exploration of the American view begins with the early pioneers' attitudes toward westward expansion when the wilderness was seen as ungodly, possessing obstacles to overcome—the Indians, wild animals, and spiritual temptation. At that time, a pastoral existence—the rural, practical, controlled state of nature—was the objective of the settlers as well as the Creator (Nash 1982).

European romanticism of the eighteenth and early nineteenth centuries slowly pervaded American perceptions of the wilderness (Nash 1982). Grounded in the astronomical and physical discoveries of the Enlightenment, natural settings could now be dignified and exalted as the handiwork of God. "Spiritual truths emerged most forcefully from the uninhabited landscape, whereas in cities and rural countryside man's works were superimposed on those of God" (Nash 1982, 46). For the first time in the American mind, particularly among educated and artistic city-dwellers, wilderness was appealing and offered an alternative to mundane matters or, worse, to the increasingly civilized society. In contrast to the pioneer experience, the wilderness had become a novelty to many, a temporary escape from the city or farm.

Yi-Fu Tuan (1971; 1974) restates the basic antithesis between the city and the wilderness and recognizes a third category—the countryside, or "middle-landscape"—where Americans such as Thomas Jefferson idealized the well-ordered garden and could avoid the increasing chaos of the cities as well as the lasting hazards (evils) of the wilderness. Tuan (1971) diagrammatically traces the transformation of cultural perceptions regarding these three categories. "Edenic" gardens (agrarian villages, middle landscapes) and early cities are, at first, perceived as sacred while the wilderness represents the profane. Eventually, the middle landscape becomes the Jeffersonian ideal with both the cities and the wilderness as the profane. And most recently, as the threatened wilderness (where recreation and education can be pursued) becomes the environmental ideal, the urban/suburban sprawl becomes the cultural "wilderness."

The recognition of the uniqueness of the remaining American wilderness landscapes reinforced postindependence nationalism, and wilderness came to be seen as a cultural and moral resource (Nash 1982). Soon thereafter, this appreciation of the wilderness resulted in the concern for its impending disappearance and the call for preservation measures. The recent preservation movement is depicted as largely rejecting the adversary relationship between wilderness and civilization (Nash 1982). Current political strategies encourage the maintenance of wilderness as an asset to the broader intellectual, spiritual and aesthetic values of American civilization.

The work of American artists and designers who attempt to simulate, mimic or re-create natural settings in urban environments reflects these values. For example, artists, architects, and landscape architects have created physical abstractions of wilderness at four sites in Manhattan, the most densely urbanized area in our densest and perhaps most chaotic metropolis. These parks and art works exploit the clarity in the juxtaposition of two pure aesthetic opposites: the wilderness and the city. By perceiving the natural landscape as a precious commodity and the urban landscape as a chaotic and uncontrollable wilderness, they reverse the scenario of monumental structures set in the pristine wilderness.

Wilderness references in the urban landscape are used for aesthetic effect, to achieve social and behavioral aims, and for narrative value. The landscapes discussed are man-made; they have been deliberately created and reflect the values of our urban culture and history in general, and of the artist and designer in particular.

The most obvious illustration is Central Park. The origins of the design for Central Park by Frederick Law Olmsted and Calvert Vaux reflect an ideological reaction or "antidote" to the negative associations of urban life (Cranz 1982). Central Park, the first of its kind, formed a man-made "naturalistic landscape" mimicking yet stylizing the indigenous environment. Olmsted and Vaux established the style and principles for park design that they and others later replicated at numerous parks across the country.

Central Park was designed as an idealized version of nature influenced by nineteenth-century European romanticism, Ralph Waldo Emerson and other American transcendentalists, and the Hudson River School of painters, including Thomas Cole and Frederick Church. These artists and philosophers "created a cult of natural beauty around such (nearby) scenic spots as the Hudson Valley, the Catskill Mountains and Niagara Falls" (Fitch 1987, 7). In addition to his work with urban parks, Frederick Law Olmsted actively supported the preservation of Niagara Falls, the Yosemite Valley, and other natural areas (Roper 1973).

Prior to its construction, the 843-acre location of Central Park at the rugged northern fringe of New York City was mostly barren land with low-lying swamps. It was inhabited by squatters and littered with refuse. Cranz (1982) points out that when Central Park was conceived in the 1850s, its concept was originally endorsed by citizenry not as a feature of the city but as a means of checking its encroachment. So, although all of Central Park was designed by man and shaped by machines, many park users perceived it as being natural.

Throughout the Park, Olmsted and Vaux utilized the main features of the natural landscape as the framework for their design. The original site possessed the principal elements of the park, including grassy meadows, parklands with high shade and little understory, and woodlands with thickets of native undergrowth. Similarly it possessed natural marshes from which to develop picturesque lakes and functioning streams (Fitch 1987).

Olmsted and Vaux felt that a pure wilderness would provide the sharpest contrast with the urbanization of New York, but the "picturesque" design was the practical compromise (Cranz 1982). This compromise, however, included large areas which were clearly imitations of natural forest environments, i.e., stylized versions of the local woodlands (Rogers 1987). The wooded areas of Central Park exhibit many of the natural characteristics of the local climax forest although they include some nonindigenous plantings and are more densely planted in the understory than would be found in a natural woodland interior. Olmsted and Vaux designed The Ramble, for example, to be seen from the distance and to be experienced from within as the wildest areas in contrast to the more stylized, pastoral settings. The Promontory, located adjacent to the 59th Street Pond, now a secured bird sanctuary, was designed as a backdrop without paths for access. Recent studies (Rogers 1987) demonstrate that although the scenic effects of the woodlands are quite striking and successful, the proliferation of self-seeded trees, which are hardier in the stressed environment of Central Park, have crowded out competing vegetation and produced significant soil erosion. So areas designed to look the most natural require extensive maintenance of their artificial ecosystems. Overuse by the public also plays a large part in the destruction of the Central Park environment.

As mentioned, Central Park provides a sanctuary for local and migratory birds. Located in the middle of the Atlantic flyway, its woodland areas in particular act as an oasis for migratory birds (Rogers 1987). Thus, the Audubon Society has become a vocal political constituency, challenging maintenance policies regarding vegetation removal in overgrown areas—those providing the greatest attraction and protection for wildlife. In so doing, they enter the current controversy in park management that questions the degree to which natural forces should be allowed to change the character and botanical composition of the park. To preserve the romanticized illusion of nature created by Olmsted and Vaux requires an active battle against the natural and man-induced pressures on the park.

Although seventeenth-century indigenous wildlife species were gone by the time the park was being developed, the original design of the park included several animals as components of the landscape, much like the installation of shrubs and trees. Swans, pea fowl, guinea fowl, English sparrows, deer, and rabbits were the wild inhabitants. Recent wildlife inventory identified five species of bats, the woodchuck, the gray squirrel, the Eastern cottontail, the raccoon, Norway rat, muskrat, several species of frogs, turtles, snakes, and nine species of fish (Rogers 1987). And as the hardiest plants in the urban woodland proliferated, so too did the hardiest animals.

The design of Central Park enunciated more than aesthetic values. Olmsted and Vaux intended that the form and function of the park be united in a popular, democratic environment—a place for rest, relaxation, and recreation for the urban masses. The design of Central Park was, thus, "undertaken as a bold democratic experiment in which all social classes were invited to mingle in scenic surroundings of uplifting poetic beauty" (Rogers 1987, 12). This early democratic ideal of Central Park parallels the contemporary attitudes toward experiencing wilderness areas as an opportunity to be stripped of worldly values and emancipated from everyday routines of the city. Interestingly, a recent user survey for Central Park clearly demonstrated that over eighty percent of the people come to the park to engage in passive activities such as reading, people watching, thinking, sunbathing, and wandering, rather than in sports, including ice skating and rowing (Rogers 1987).

Relating Central Park to wilderness values involves the educational value of the visitor's experience. Some of this education comes through subtle observation. Some comes from more structured educational opportunities in the form of the bird sanctuaries, nature walks, information centers, and various publications.

Psychologically, the park creates an escape from the geometric order of the streets of Manhattan, from the architectural jumble, from the noise and frantic movement of traffic on city streets. A green island within the island of Manhattan, it represents an image of a benign and domesticated wilderness. At night, however, the park reinforces the early American view of wilderness as a place of doubt and foreboding. Central Park becomes a negative space: Manhattan's landscape of fear (Tuan 1979).

On the design of the Ford Foundation building, urban designer and critic Jonathan Barnett (1968, 112) writes, "In a time of rapidly increasing urbanization, growing densities, and a more frantic pace of life, a garden court of this kind becomes an oasis of tranquility and a point of balance in an uncertain world." Similar to the premise for the design of Central Park about a hundred years earlier, the architectural firm of Roche and Dinkeloo in their design for the Ford Foundation utilized a wilderness reference to elicit a sense of sanctuary and escape from the harsh and uncontrollable aspects of the urban environment. Also, like Olmsted and Vaux's work in Central Park, this project served as an archetype for a building form—an office building with a central landscaped atrium—which is now almost commonplace, but was revolutionary for its time.

The Ford Foundation courtyard defines an intermediary space between the sealed environment of an office building and the aesthetic and social reality of city life. To the office worker this idealized and highly controlled outside contrasts with, and perhaps provides a barrier to, the real outside. Offices have sliding windows which open onto the court, permitting the occupants to participate in the common space. The regulation of the temperature in the court reinforces the sense of relief from the harsh outer environment, ranging from a maximum of 45 degrees in the winter to 90 degrees in the summer (Burns and Smith 1968). On a winter day, in particular, this lush, green environment contrasts with the gray, leafless trees in the park across the street. Although the wilderness allusion in this highly artificial landscape is purely aesthetic and sensory, two small signs provide a note of ecological reality concerning the collection and utilization of natural rainfall.

Various points of view present playful ambiguities. Looking into the glass wall of the building from the hard surfaces of the street, we almost seem to be looking out into this wooded landscape. What continues to set this landscaped atrium apart from many that were built later is that landscape architect Dan Kiley used plants that would simulate a natural environment, that would seem familiar to New Yorkers even though the plants were not native to the region. More recently landscaped atriums in New York typically contain more exotic flora.

Urban references to the wilderness appear to increase seasonally each December. Origins of the use of evergreens as Christmas icons are traced to European paganism (Miles 1912; Snyder 1976). Imported evergreen trees redirect our attention from the dormant street and park trees and reinforce the notion of an eventual return of Spring.

Located in the Soho district of lower Manhattan, the Time Landscape by Alan Sonfist is a reaffirmation of his childhood experiences in the Bronx Park area which then contained the last virgin forest in New York City. Sonfist, an environmental artist, focuses on natural systems which have been destroyed and replaced. He points out that public monuments normally celebrate historical events. Therefore he offers the Time Landscape as an alternative kind of monument, one which documents "the transformation of natural, non-human phenomena" (Lippard 1981, 149). Sonfist's Time Landscape reinforces the comparative stature of preserving the natural environment as part of the city's history along with the architectural environment. The Time Landscape attempts to remind us of the diminution of unspoiled space in the city and on the planet, and the reclamation of such environments.

Sonfist based the Time Landscape, established in the late 1970s, on early settlers' accounts of the local geology and biology and the team research of a biologist, a geologist, an ecologist, and an historian. It is planted with only native species, some of which were reintroduced to Manhattan. The piece attempts to simulate the natural process of ecological succession beginning with grasses and wildflowers, followed by pioneer species such as red cedar, and culminating with native dogwood, beech, oak, sweet gum, tulip trees, and maples (Nadelman 1979). Thus, the observer can experience, while looking behind the tall wrought iron fencing, "an image of precolonial land in the midst of the colonized and exploited urban site" (Lippard 1981, 149). In effect, the landscape serves as an illusion in time; Sonfist has replicated the environment of a hundred years of growth in a matter of months. Perlberg's (1978, 84) review of this sculpture reasonably states that "Sonfist's forest can be understood as conceptual myth-making, imitating a process disturbed by man only to be memorialized by him."

Unlike Central Park and the Ford Foundation, Time Landscape does not simulate a wilderness setting as a background for human activity. The fence surrounding it assures that it cannot be entered. Thus, this environment more literally meets a dictionary definition of wilderness as "uninhabited by human beings" (Webster's Third New International Dictionary, 1986 printing, 2615). Once again, however, local residents, complaining that this corner is dark and dangerous at night, fear the landscape.

Sonfist, in bringing attention to the subject of wilderness, presents a public message. In his own words, this landscape reinforces values inherent in many of our wildernesses. "I think that we have to understand our own heritage and that we are animals that come from a forest type of structure. Time Landscape is a reminder of our past, of our basic selves and of our total relationship to the cosmos" (in Herrera 1977, 54).

The effect of the piece today is quite different from the artist's original intent. Opportunistic invasive plant materials have turned the site into one more reminiscent of an overgrown abandoned lot than a virgin forest. It appears more an unruly component of the chaotic urban fabric than a purist contrast to it. Without vigilant management a small patch of land in a densely urbanized setting cannot be sustained as a native northeastern forest. The plants in the Time Landscape are, however, characteristic of the temperate zone urban ecosystem, and many have been present in urban areas since ancient times.

In a recent collaboration, artist Mary Miss, landscape architect Susan Child, and architect Stan Eckstut, invoked a somewhat more abstract wilderness representation. On a massive landfill site, developed with high rise office and residential buildings in New York's newest district, Battery Park City, they designed one piece, referred to as South Cove, of a long continuous waterfront park. In contrast to the other, extremely formal portions of the promenade, the creators decided that "nature should play a dominant role" (Karson 1986, 50).

Susan Child states, "A cove is a place of repose, protected, fortified by a strong grade, and strong vegetation at the mouth. We decided to adopt this scheme for our cove, and to try to emphasize these special qualities" (in Karson 1986, 50). The park consists of three zones: an upper level which is formal, in keeping with the character of the city beyond and adjacent portions of the promenade; the intermediary level which is less structured; and the heavily planted point at the southern end of the cove which suggests a wild habitat and is planted with indigenous shore species including beach grass, rugosa rose and Japanese pine (Karson 1986). Throughout the piece, the contrast between natural forms and native plant materials and between man-made forms and elements is emphasized.

Although South Cove was created on an entirely artificial land mass, as in Sonfist's work there are references to the pre-developed state of the island of Manhattan. Rocks at the base of the planted slope recall the character of an original shoreline. Some of these rocks disappear into the pavement, as they would have into the river. Areas densely planted with indigenous species are inaccessible, again recalling an uninhabited wilderness landscape.

Like the landscapes discussed previously, at South Cove the artists and designers felt that the recollection of a natural environment would be therapeutic for the urban public. Artist Mary Miss explains, "I want South Cove to ameliorate the harsh experience of the built environment" (in Karson 1986, 50). Though the work romantically represents a natural landscape, it also functions as a realistic urban landscape. Rather than emphasize the conflict between the two environments, the piece suggests coexistence.

Modern urban culture has idealized the American wilderness. This has resulted in attempts to simulate primarily its benevolent connotation, in sometimes incongruous contexts. Allusions to the wilderness in the urban and suburban landscape are commonplace. In these examples, the allusion was essential to the aesthetic and narrative intent of the work. The artist or designer instills societal values in his interpretive re-creations of wilderness; perhaps these works, in turn, reinforce these values in the minds of those who enter and observe them.

There is a new movement in our culture towards recognizing the existence of the natural environment within the city. The field of urban forestry has emerged both in academia and in practice as an outgrowth of that trend. It deals specifically with many of the issues raised by the coexistence of natural systems and the city, such as the dilemmas involved in regulating the wooded areas of Central Park and the Time Landscape. Landscape architects are also beginning to focus on ecological principles, even in projects located in the most dense urban settings such as the South Cove in Battery Park City.

Those in American cities evidence a growing interest in the conservation of true wilderness areas. In contrast to the landscapes deliberately designed for human use based on the aesthetics of our culture, in these places people can observe an environment created by natural forces. However, to accommodate the public, these landscapes, too, must be actively managed. These urban wildlands serve to ease the psychological burden of overdevelopment, provide ecological benefits such as cleaner air and preservation of wildlife habitats, and provide accessible wilderness for recreational and education purposes.

A sizeable sector of American society has no access to, or experience with, remote wilderness areas. Therefore, both urban wilderness areas and the artistic representations of wilderness in cities in some ways serve as more democratic retreats from city life than traditional wilderness. It is difficult to say whether these landscapes serve to instill respect for wilderness; they usually provide less challenging experiences and little solitude. However, the appreciation of trees, water, and clean air, and the absence of cars and other signs of industrialized society seem universal and are undoubtedly reinforced by these experiences.

References

Barnett, J. Innovation and symbolism on 42nd street. Architectural Record 143(2):105–112; 1968.

Burns, J. T.; Smith, C. R. Charity begins at home. Progressive Architecture 49(1):92–105; 1968.

Cranz, G. The politics of park design. Cambridge, MA: MIT Press; 1982.

Fitch, J. M. Central Park: A paradigm for socially useful landscapes. In: Rogers, E. B., ed. Rebuilding Central Park. Cambridge, MA: MIT Press; 1987:7–9.

Herrera, H. Manhattan seven. Art in America 65(4):53–54; 1977.

Karson, R. South cove. Landscape Architecture 76(3):48–53; 1986.

Kostof, S. A history of architecture. Oxford, England: Oxford University Press; 1985.

Lippard, L. R. Gardens, some metaphors for public art. Art in America 69(9):136–150; 1981.

Miles, C. A. Christmas in ritual and tradition Christian and pagan. London: Adelphi Terrace; 1912.

Nadelman, C. What's that forest doing in Greenwich Village? Artnews 78(9):66–71; 1979.

Nash, R. Wilderness and the American mind. 3d ed. New Haven, CT: Yale University Press; 1982.

Perlberg, D. Reviews. Artforum 17(1):84; 1978.

Roper, L. W. A biography of Frederick Law Olmsted. Baltimore: The Johns Hopkins University Press; 1973.

Rogers, E. B. Rebuilding Central Park. Cambridge, MA: The MIT Press; 1987.

Snyder, P. V. The Christmas tree book. New York: The Viking Press; 1976.

Tuan, Y. F. Man and nature. Association of American Geographers, Resource Paper No. 10. Washington, DC: Commission on College Geography; 1971.

———. Topophilia: A study of environmental perception, attitudes and values. Englewood Cliffs, NJ: Prentice-Hall, Inc.; 1974.

———. Landscapes of fear. New York: Pantheon Books; 1979.

Weinstein, M. A. The wilderness and the city. Amherst: The University of Massachusetts Press; 1982.

Wilderness as a Human Landscape

STEVEN R. SIMMS

D A W N broke still and hot in the wadi bottom where we had camped among canyons three thousand feet deep. In the rising light, the brightening pastels unfolded on the plain before us. A desert wilderness of stark beauty, seemingly unchanging with the strength that wilderness conveys, assures the power of nature over man. We were anthropologists and had hiked for two days in searing heat, over slickrock, along ragged desert escarpments towering thousands of feet above our destination below. We descended tumbling staircases of narrowing wadis to reach this place. This calm, Near Eastern place that produced prophets in biblical times. This was the ancient Edom of the Old Testament—present-day central Jordan.

First light revealed the colors of the sandstone walls, the black schist of the scarp and a nearby low-sculpted bench of gravels left by the ancient stream bed we had chosen for a camp. I reached across my sleeping blanket and ran my fingers through sands still warm from the hot night. We are truly remote, I thought, and this *is* the Arabian outback! But above and somewhere hidden within the profile of the gravel bench came sounds. At first, it sounded like the distant cries of a small animal in distress. But the noise was jumbled by the rococo terrain, and listening always changes sounds. It was the on, off, haphazard cacophony of goats in a herd. But what was that other sound? That singsong, rhythmic ebb and flow of shouts and murmurs, cackles and whoops? It sounded human. I climbed the small cliff behind our camp. Rising to the lip where the flat top of the gravel bench was visible, I was startled by the sight of two *Bayt Shar*, the long, black, goat-hair tents of the Bedouin. The goats I had heard so muffled and confused were high on the cliffs above. To the songs and calls of a young Bedouin herder, a member of the *Sa'idiyin* tribe I later learned, they had begun their daily search for forage in a landscape so barren as to defy reason to such an economy. It seemed irrational. It seemed out of place. It must be human.

Of course, I thought. These people have stripped this landscape of its vegetation. They overgraze to pursue their way. The area around the tents was littered with the mark of humanity; charcoal, ash, trash, and goat dung were everywhere. I wondered how they continued to find forage for their goats in such a place. This was no wilderness. They had destroyed it. As I descended the cliff to our camp, set at the edge of the Wadi Araba, the famous Dead Sea Valley, I had to smile at the irony.

These people are altering their landscape. Archaeology shows their ancestors have done it this way, "traditionally," for thousands of years. They did it in other ways for tens of thousands of years before that. Perhaps it is a human landscape—a human wilderness. Yes, these simple people were destroying the wilderness with their simple, outdated ways. I wondered if doing that traditionally, as part of their simple culture, and for a very long time, gave them any right.

Southern Utah. What a wonderful wilderness—that of it left. Utah, a state so bent on economic development and short-term revenues that it has less designated wilderness than any other western state. Here is a vast landscape of sandy-bottomed canyons, winding staircases of natural sandstone—domes, fins, towers, and all those nipples so eagerly described by Ed Abbey years ago. All this liberally bedecked with towering mountains! Despite the effects of our shameless development marching over the American landscape, one can still find solitude in southern Utah. No people.

According to modern conception, people typically taint the pristine nature of wilderness. We envision a wilderness without people, granting them a place only in romantic, primitive nations. Awe inspiring as such a retrospective might be, this conception is based on mistaken images of noble savages as the only people to ever completely harmonize with nature. Consistent with our view of a wilderness in harmony, we necessarily exclude people, or at least all the people who are not noble savages.

Imagine a scene in southern Utah set in a remote canyon over 700 years ago. It is one of those warm August evenings, deceptively darkened by an approaching thunderstorm. An alcove above a ledge overlooks the canyon. There is a buzz of activity. The smell of humanity lumbers down the canyon in the humid air of the storm. Aromas of human, turkey, and dog concentrate in a sweltering mass defined by the cramped living space of the canyon walls. A smoky haze rises from small holes in the roofs of the stacked masonry cubicles and mingles down canyon with the ever-present ordure. Perhaps twenty people can be seen on the rooftops and in front of the alcove. Ten more can be seen in the canyon below, for no tree or shrub is in immediate sight. They have been cut for construction materials, and for firewood to provide cooking fuel and heat during the cold winters. Five Indians file down the steep trail with the nimbleness of long familiarity. They have lived in the canyon country all their lives as have their parents and grandparents. They are one with nature. Their world view exemplifies this—no notion of wilderness, no separation of humanity and nature. No juxtaposition of humanity *versus* nature. They are Anasazi Indians. Does an ideology ensure harmony between behavior and nature? Perhaps not.

The men come from corn fields where they have cleared the shrubs and timber to farm (Lister 1966; Stiger 1979; Betancourt and Van Devender 1981). They hunt for prime-age, sometimes pregnant, female mule deer, favored because they carry more fat and yield hides unblemished by the fights common among males. In recent years, some of the herds have been depleted, a combination of overhunting and depressed fertility rates caused by too much reliance on the favored females. Some of the mesas offer so little game that the women must break up the bones to boil for the last remaining nutrients and for the ever-needed fat. They didn't always have to do that, and the women don't like all the extra work. Stories told around the hearth capture the golden days of plenty— the past. Always the past. An old man begins to counter, "The past wasn't always great. Let me tell you about the time." His words trail off as the group favors the nostalgia that salves their current problems.

Over a temporal span that outstrips the limits of oral history, the Anasazi's human wilderness was repeatedly overtaxed by the precarious relationship between people, agriculture, and drought, all in the context of a fragile landscape (e.g., Schlanger 1988). The interactions of the Anasazi with their environment produced a spatiotemporal mosaic of negative impacts, harmony, temporary balance, and imbalance.

The people at the village hope that the storm will bring just enough precious rain, but not so much that it washes away the crops they have planted behind masonry barrage dams in the canyon bottoms. They used to not worry about this, but they did not always try to control the canyons. Only after the limited fields on the mesa tops were appropriated and population continued to grow, did they construct canyon terraces to create more land (e.g., Winter 1977). A purchase of insurance? Environmental impacts?

That night in their ceremonial lodges called kivas, the men performed rituals to help bring rain. By the firelight of the late summer evening, the entire village hoped that the ritual power would work. They all hoped there would be enough food for the winter. They have *always* worried about that. It is the year AD 1268 in the southeastern Utah wilderness, late in the history of the Anasazi (e.g., DeBloois and Green 1978; Matson and Lipe 1978; Cordell 1984). The people are trying to eke out a living from a landscape shaped by both natural forces and human hands. For hundreds of years, repeated group fission, local abandonments, migrations, and recolonizations characterized the Anasazi settlement system (e.g., Matson and Lipe 1978; Cordell 1984; Schlanger 1988). They made adjustments to a changing natural world that they significantly shaped.

The Anasazi had done it this way for over a thousand years. Their ancestors had used the land in other ways since colonizing the Americas over ten millennia earlier. Some of the impacts of their agriculture on the canyons and of clearing the land remain with us even today. Perhaps it is a human landscape, a human wilderness.

These images from the modern Near East and ancient southern Utah suggest what an anthropologist might see in a wilderness. Anthropology, the science of humanity in the most comprehensive sense, always focuses on the human. This creates a curious juxtaposition when the subject is wilderness. Could it be possible that wilderness, by some current American definitions, does not include humans? Language and definitions mirror the cultural template that textures our lives and gives meaning to our experiences. In a world such as ours, where land is set aside from the march of development and managed by huge bureaucracies often amidst political firestorms, perhaps wilderness cannot conceptually include humans.

This idea can be explored from the large temporal and cultural scale offered by anthropology. Although some may define wilderness as a world of the distant past without people, the hand of humanity has shaped the ecosystems of the western hemisphere for a minimum of 12,000 years and the remainder of the planet for many tens of thousands of years before that (e.g., Fagan 1986). Perhaps the truly pristine is restricted to times prior to human evolution; but it may be as significant as it is amusing to note that humans are the only animal concerned with issues such as wilderness and the pristine. Perhaps any wilderness created today may be less a matter of restoring nature to her pristine form than the creation of hypothetical ecosystems approximating our vision of the distant, prehominid past. In actual practice, most conceptualizations of wilderness probably are restricted to relatively recent visions of prehistoric times. The definition is far from intrinsic, and the plasticity in definition is relevant to an understanding of wilderness.

One ingredient in the popular conceptualization of a wilderness without humans adheres to the eighteenth and nineteenth century notions of progress. That is, progress from a pristine state of nature, "noble savagery," into a "civilized" condition where nature is perceived as tameable. Progress from a humanity that was ostensibly in harmony with nature to one now in a shambles of disharmony. A substantial literature has argued for a variety of positions as to American Indian attitudes toward conservation and the natural world (see White 1984 or Callicott 1989 for reviews). Writings exhibiting diverse conclusions typically have valid points embedded within them. Indians have been seen as utilitarian conservationists (e.g., McCleod 1936; Speck 1939), as rational man subject to incentives (Baden, Stroup, and Thurmon, 1981), through the negative lens of ethnocentrism (Guthrie 1971), and through the favorable lens of nostalgia (e.g., Nabokov 1978). The increased environmental consciousness of the 1960s and 1970s and the concomitant increases in the popularity of Indians led to extreme expressions of yet another stereotype—this time one of pristine, precontact Indian people living in static harmony with all aspects of nature (Strickland 1970; Jacobs 1972). As popular and as politically useful as this notion continues to be, it is well established that simple societies nevertheless have significantly shaped their environment (e.g., White 1980; Dobyns 1981; Martin 1981; Cronon 1983; Diamond 1988).

Previous positions have tended to describe simple societies through normative characterizations. These tend to stereotype an entire society, discounting that the most significant topic of study might be *variability* in behavior and attitudes. However, behaviors and values about the natural environment are highly variable both spatially and temporally within cultures, as well as between them. In this behavioral context, conditions of selection that favor or inhibit conservation and related ethics will occur through time. The conditions of selection need not be rational in the sense of formal economics since forces favoring or hindering an "environmental ethic" can arise from many quarters. However, until variability, rather than stereotypes of simple societies is described, there will be no basis to seek processual explanations for cultural form and change. Until variability becomes the subject matter, not merely an inconvenient obstacle to characterization, the goal of explanation will remain stunted by descriptive work emphasizing one normative characterization or another.

We are just beginning to understand the role of hunter-gatherers and other simple societies as potentially significant predators as worthy of study as the better known wolves, coyotes, and raptors (Nudds 1987). Since all human societies have played a role in shaping ecosystems, the evidence suggests that simple societies may have caused as significant an environmental damage as have more complex societies. To be sure, there is a difference of scale, with complex societies altering nature on an increasingly global scale. On the other hand, there is nothing intrinsic to hunter-gatherer, band society that ensures harmony with nature. In fact, there are many cases of hunter-gatherer behavior that, when judged from a pro-wilderness stance of today, would be classified as having negative impacts upon wilderness environments.

One example of how hunter-gatherers participate in shaping ecosystems is the effect North American Indians have had on the large-game populations of our continent. At one extreme, there was the waste of dozens of bison carcasses at the Olsen-Chubbuck site on the high plains more than 8,200 years ago (Wheat 1972; Frison 1978). After driving a herd of bison over the cliffs, only the upper layers were butchered, leaving many to rot. Ironically, the greatest selectivity in butchering and the absence of complete processing for storage occurs in the earliest sites, when the New World was "pristine," and declined somewhat through time (Todd 1987; Kelly and Todd 1988). Thus, while all plains bison kills do not exhibit such graphic waste as seen at Olsen-Chubbuck, the realities of hunting, with associated decisions regarding butchering, transport, and storage fostered a departure from the popular notion that Indians, *under all circumstances,* ate everything and took only what was "needed." After all, what constitutes "all" of the animal? Is it only the cuts of meat that would please a modern American, or would it include viscera, bone marrow, and bone grease too?

Hunters in simple societies appear to so overhunt in the vicinity of their villages that in cases such as the Yanomamo of Venezuela, game is depauperate in those areas (Chagnon and Hames 1979). The Iroquois and other Native Americans of the northeastern United States regularly and extensively burned away the forests to improve the hunting (Pyne 1982). Apparently they did not practice a conservation ethic of traveling to distant hunting grounds before reducing local game densities. Or perhaps they were not conservationists simply because, as in the case of the Yanomamo, their neighbors would not let them travel to distant hunting grounds (Chagnon and Hames 1979).

Hunting large game created a tremendous and widespread impact in North American prehistory, especially in the western United States. As nonhuman causes, hunting, or interactions of these forces depleted the game, hunter-gatherers added alternative and less desirable resources to their diet. Some of the most diverse diets in the world were once found in the Great Basin of the western United States, yet ethnographies show that large game species, rarely tabooed, were the favored resources (e.g., Steward 1938, 1941, 1943; Stewart 1941, 1942). The presence of broad, but fluctuating, diets in places like the Great Basin shows that hunter-gatherers participated in shaping the structure of resources in their environments (see Simms 1985, 1987). At the extreme, in some times and places, such foraging strategies may have maintained animals such as elk or deer at levels we might declare worthy of endangered species status today.

Archaeologists have argued for decades about whether the human colonization of North America actually caused the extinction of large, Pleistocene herbivores such as the woolly mammoth or giant ground sloth (e.g., Martin 1967). The debate has now moved beyond the simplistic dichotomy blaming *either* the natives *or* the climate. Current efforts to understand the Pleistocene extinctions center around modeling the influences of humans interacting with other variables (see Meade and Meltzer 1985 or Grayson 1987), tacitly acknowledging that humans in simple societies play a role in shaping ecosystems that can be significant without being catastrophic.

The hypothesis that Native Americans were not intrinsically and inevitably interested in conserving large game populations is further attested by the archaeological record of hunting. Faunal records from archaeological sites can sometimes yield evidence for the age and sex structure of hunted animal populations. Unlike nonhuman predators, which tend to hunt the old, sick and the young, humans can take all kinds and are capable of greater selectivity. While human predation is variable, the hunting of bison on the plains shows a bias toward females with young, and usually animals of prime reproductive age, rather than the old and infirm (Frison 1978). At Sudden Shelter, a hunting site in Utah spanning the period from 8,000–3,000 years ago (Jennings, Schroedl, and Holmer 1980), twenty-five percent of the mule deer hunted were juveniles or even prenatal, suggesting the hunting of pregnant females (Lucius and Colville 1980). At Gatecliff Shelter, in Nevada, the hunting of bighorn sheep (the focal activity at the site) was biased toward prime-age adults, probably both males and females (Thomas and Mayer 1983).

Hunter-gatherers may need to hunt all year, not just when game is "in season" as in modern times. Thus, the short-term need to feed people can decrease the ability to routinely consider what is best for game populations. Certainly at some level, it is in the hunters' best interest to be concerned with the welfare of game populations. But even here, there is much room for variability. If hunters could readily move to new ranges, the constraints against overhunting would be less than if neighbors or the costs and risks of abandoning other resources made such moves costly. Thus, different behaviors by hunter-gatherers would have a significant impact on the structure of game populations.

Hunter-gatherers also consider different variables when hunting than modern Americans do. For example, hunting females of prime age makes sense from a nutritional point of view since females are fatter (Speth and Spielman 1983). One would expect precisely the opposite behavior if conservation of game populations was of *exclusive* concern. This is not to argue that hunters routinely and purposefully degrade their environment, but to show the relevance of variables in aboriginal hunting that are not often considered.

Given the circumstances of aboriginal hunting, the familiar, and stereotyped Native American ideologies encouraging hunters to only take what is "needed" would be consistent with an ecological context of low game densities shaped by human impact. The point is not to challenge the veracity of Native American ideology, but to suggest that the concept of the pristine is not its inevitable corollary.

Perhaps some of our views of Native Americans have been distorted by explorer and traveler accounts that came with the relatively late westward colonization of the white man. Interestingly, the extermination of Native Americans was not simply a function of the well-known, face-to-face atrocities of Euro-American society. Evidence shows that "their numbers became thinned" on a continental scale from epidemic diseases brought to Florida by the earliest Spanish explorers during the early 1500s (Dobyns 1983; but also see Crosby 1972; Ramenofsky 1987; and Thornton 1987). Wave after wave of smallpox, measles, influenza, bubonic plague, diphtheria, typhus, cholera, scarlet fever, and other diseases reduced the precontact native population by estimates as high as seventy-five percent during the sixteenth century (Dobyns 1983). Since the epidemics were intermittent, Native Americans could not achieve widespread immunity from diseases such as smallpox, the major killer. This enabled over three dozen documentable smallpox epidemics to spread across the continent between 1520 and 1898 (Dobyns 1983). Consequently, these diseases could have spread nationwide with catastrophic effects over 200 years before the actual westward migration of the white man (Dobyns 1983). As the demographer Henry Dobyns has said, "The North American continent was thus in one sense not the 'virgin' land many historians and politicians have called it. It became 'widowed' land by the time of widespread Euroamerican settlement" (Dobyns 1983, 8).

By the time the early explorers such as Lewis and Clark arrived in the West at the beginning of the nineteenth century, Native American populations would have been so reduced and disrupted that game could have actually returned to some areas in atypically large numbers. Could it be that the abundance of game that awed early explorers was already an artificial situation brought about by the massive depopulation of a continent, thus eliminating a major predator from the natural ecosystems?

Kay (1987, 1990), in studying the role of Native Americans in shaping large animal populations in North America and on elk populations in Yellowstone National Park, suggests that this may have been the case. Accounts by the earliest explorers emphasizing the abundance of game are further placed in historical perspective by later accounts during the white colonization of the western United States attesting to the paucity of large game. Early depopulation would have been followed by migrations and local repopulation, creating a dynamic mosaic of aboriginal foraging groups responding to the combined effects of introduced disease and westward colonization by Euro-Americans. In turn, game abundance in the United States would also have been highly variable from region to region—not consistent with a scenario of continental superabundance envisioned by wilderness romanticists.

The archaeological and ethnographic records also suggest that large game were not superabundant in prehistoric North America. In fact, much of the desert West appears to have been a highly variable patchwork of game abundances, allowing extremely broad, Native American diets. Rather than the result of environmental consciousness or wilderness values, diets focusing on game, under many ecological circumstances, would have been the most competitive adaptive strategy possible (Simms 1987). If humans have been a significant predator for the 12,000 or more years Native Americans have occupied our continent, then it follows that wilderness has been artificial, hence subject to varying definitions since the extermination of these people.

Besides directly affecting game distribution and density, humans in simple societies have actively shaped their environments in other ways. A more pervasive influence is in the aboriginal use of fire, an agent affecting entire ecosystems (Lewis 1973; Martin 1973; see Pyne 1982 for an excellent overview). Native Americans used fire as a hunting aid, as a means of creating hunting zones, in warfare, to aid various forms of plant collecting, to simplify wood collecting, and to reduce insect pests (Pyne 1982). Massive evidence shows that Native Americans modified the American landscape with fire on a continental scale with the effects most broadly seen in the creation of grasslands. The accidental and intentional scorching of millions upon millions of acres of forest and shrubland by Native Americans expanded grasslands and barrens in virtually all areas of the continent (Martin 1973; Pyne 1982). In fact, the historic suppression of fire has created more forests today than had graced the continent since aboriginal colonization many millennia ago. This too challenges the concept of wilderness as a temple to nature's harmony, free of the unnatural intrusions of man. In his book *Fire in America*, Stephen Pyne (1982) writes, "Because the concept of wilderness has been linked to the discovery and settlement of America by Europe, any enterprise that predates the reclamation, no matter how extensive, is considered wild. Thus, Indian fire practices, which were enormously powerful as landscape modifiers, have been dismissed. The return of natural fire to wildlands is less likely to 'restore' an ancient landscape than it is to fashion a landscape that has never before existed" (Pyne 1982, 17–18).

The mountains, canyons, and deserts of the Southwest are among the greatest treasures of American wilderness. Modification of the wilderness by ancient societies there can be seen in the deforestation and soil erosion most graphically represented by the Anasazi societies of the Chaco Canyon area (Betancourt and Van Devender 1981; Betancourt, Dean, and Hull 1986). For example, Betancourt, Dean, and Hull (1986) argue that logging activities in the Chaco Canyon area of northwestern New Mexico in Anasazi times had significant environmental effects. These activities did not necessarily eliminate forests, but were similar to the effects of timber-thinning techniques. The Mesa Verde area provides evidence of purposeful manipulation of timber sources (Nichols and Smith 1965). Landscapes such as these would hardly pass as wilderness by the standards of today given the notable incongruity between the interests of wilderness and the timber industry.

Other examples of the impact of simple societies around the world include the extinction of hundreds of species of animals by the aboriginal colonization of Polynesia (Diamond 1988). The extensive grasslands of east Africa which support some of the best-known, large-game populations in the world were shaped by human intervention over a minimum of 1,600 years (Feely 1980). The desertification of the Near East and North Africa resulted in part from thousands of years of human impact beginning before the domestication of plants and animals 10,000 years ago and continuing up through the ancient civilizations of the Old World. In fact, deforestation in parts of the Near East led to an environmental disaster by 5,000 B.C., readily apparent in decreased timber size in dwellings, a decrease in the use of lime plaster which required large amounts of charcoal to manufacture, and a decrease in the diversity of wild animals (Pringle 1990). Finally, the potential of even very simple societies to run counter to the noble-savage image can be seen in an act as symbolic as the cutting of an entire orange tree to obtain a few oranges by the Ache' of modern day Paraguay (K. Jones, personal communication). Yet the Ache' are one of the most simple hunting and gathering cultures remaining in the world today (e.g., Hill and Hawkes 1983; Hill et al. 1985).

We are rapidly learning from evolutionary ecology, a perspective used in anthropology, that cultural systems, any time, any place, can be seen by the participants as conservationist or abusive of their environment. It is not the level of cultural complexity, the time, the place, or the value system that matters as much as the conditions created by combinations of these. A corollary of this view, deeply embedded in anthropological theory, argues that the conditions of existence, the problems of life, tend to shape the evolution of value systems in culture as well as the degree of symmetry between ideal and actual behavior. As a research strategy, this has proven more fruitful than the reverse in which a set of indelible values mysteriously arises *sui generis* from the mind of the native. A negative influence of the dogma that culture is an immutable, self-determining entity becomes especially apparent in the case of Native Americans where a concern with political correctness and an idealist perspective on the evolution of culture continues to cloud the literature on Indian-land relationships (*cf.* Callicott 1989). Cultural context is indeed relevant, but to assume that cultural systems of ethics and values are the unchanging, primary determinants of behavior begs questions as to which factors shape variability in these very value systems. We cannot solve questions about dynamics by appeals to normative characterizations about systems of ethics and values, no matter how politically comforting they may be (see Blurton-Jones 1976; Smith and Winterhalder 1981; or O'Connell, Jones, and Simms 1982 for discussions of the differences between evolutionary ecology and other ecological perspectives common to the social sciences).

The issue goes beyond romanticizing and stereotyping our ancient ancestors to the consequences of such stereotypes. The view that hunter-gatherers are shining examples of nature's harmony before our fall from grace is saturated with predestination and fate. Ironically embodied in this romantic, noble savage view is the conclusion that we in modern society cannot, by definition, be in harmony with nature. In this myth, harmony was the exclusive property of bygone days. This static view reflects ignorance of the lessons of time. By adopting this view, all we can do is retain some token wilderness to preserve the memory of the myths we have created: that landscapes in the good old days did not contain people and that if these landscapes did contain people, they were surely an antediluvian and natural people, harmonizing with a world yet to be destroyed by civilization.

Despite the horrific tone accompanying some of these examples, the intent here is not cynical. Given the perspective of government and public policy today, we have no choice but to agree with many prowilderness lobbyists who seek to minimize the impact of people in modern wilderness. The point is not to counter this modern fact of sociopolitical life, or to argue for unbridled development, justifying our present sins by demonstrating the time depth of human impact. Rather, it is to show the danger in thinking that the pristine is attainable, that static preservation is possible, and that a universal, or even obvious, definition of wilderness is possible.

The answers to policy questions will always be difficult, but perhaps they will be easier to glimpse with a more realistic concept of wilderness. In the day-to-day drudgery of policy-making, lobbying for wilderness, and picking up other people's trash from littered lands, it is easy to align with the fatalistic notion of progress, that naively romantic notion of a simpler past when all was in beauteous harmony, that mythical view of wilderness restored to its pristine form, apart from humans.

Optimistically, evidence suggests that conservation may actually have fewer chances to occur in simple societies than under more complex circumstances (Layng 1986). When a low human population density exists, pressures favoring conservation may be weak compared to the advantages of exploitation, even if exploitation is described by the participants' ideology as "only taking what is needed." Conservation on a significant scale more likely results when people apply sanctions upon other people for environmental transgressions. Thus, the rise of a conservation ethic may be more commonly associated with the rise of more complex societies with their explicit systems of territory and ownership, backed by legal authority. While committing the most massive destruction because of our sheer magnitude, this view argues that it is our society that holds the greatest promise of a solution to planetary environmental degradation.

Furthermore, the plasticity of human attitudes toward conservation and exploitation, as demonstrated by the variance across cultures and through time, can give us hope that all is not lost. Wilderness has never been static, never been in balance, and can probably not be free of the hand of man. Wilderness will always be an artificial entity; thus we are forced to choose the kinds of artificial situations we wish to define as wilderness. We must define which "wilderness" we want, and then create conditions that favor the promotion of these artificial and dynamic situations in our own postindustrial society. A realistic assessment of our own possibilities is superior to creating nostalgic myths that only grant this goal to the primitive. Since the choice of definition is not intrinsic, or even obvious, the decision-making process is of vital importance. The choices will inevitably be made in the political context of our time and be subject to many forces. In the arena of American politics, one person's wilderness can be another person's hell.

[Editor's note: Removing adult females and juveniles in game populations can result in good harvest management, depending upon a population's circumstances. Certainly, Native Americans may not have been aware of this point.]

Acknowledgments

The views expressed here have evolved over a number of years through study of hunter-gatherer ecology. My interest was rekindled during informal talks by Charles Kay and Frederic Wagner at Utah State University in the fall of 1988. These discussions suggested the relevance of the topic to wilderness issues. Others who contributed through conversations and reviews of the manuscript include Marina Hall, David Lewis and Carol Loveland, Utah State University; Kevin Jones, Utah Division of State History; James O'Connell, University of Utah; Samuel Zeveloff, Weber State University; and two anonymous reviewers. I appreciate the insights and encouragement of these people and, of course, assume sole responsibility. This paper was read at the 1st North American Interdisciplinary Wilderness Conference, Weber State University, February

1989, and at the Mountain West Center for Regional Studies, Utah State University, April 1989.

References

Baden, J.; Stroup, R.; Thurmon, W. Myths, admonitions and rationality: The American Indian as a resource manager. Economic Inquiry 19:132–143; 1981.

Betancourt, J. L.; Dean, J. S.; Hull, H. M. Prehistoric long-distance transport of construction beams, Chaco Canyon, New Mexico. American Antiquity 51:370–375; 1986.

Betancourt, J. L.; Van Devender, T. R. Holocene vegetation in Chaco Canyon, New Mexico. Science 214:656–658; 1981.

Blurton-Jones, N. Growing points in human ethology: Another link between ethology and the social sciences? In: Bateson, P. P. G.; Hinde, R. A., eds. Growing points in ethology. Cambridge, England: Cambridge University Press; 1976:427–449.

Callicott, J. B. American Indian land wisdom? Sorting out the issues. Journal of Forest History 33:35–42; 1989.

Chagnon, N. A., and R. Hames. Protein deficiency and tribal warfare in Amazonia: New data. Science 203:10–15; 1979.

Cordell, L. S. Prehistory of the Southwest. New York: Academic Press; 1984.

Cronon, W. Changes in the land: Indians, colonists, and the ecology of New England. New York: Hill and Wang; 1983.

Crosby, A. W., Jr. The Columbian exchange: Biological and cultural consequences of 1492. Westport, CT: Greenwood Press; 1972.

DeBloois, E. I.; Green, D. F. SARG research on the Elk Ridge project, Manti-LaSal National Forest, Utah. In: Euler, R. C.; Gumerman, G. J., eds. Investigations of the Southwestern Anthropological Research Group. Flagstaff: Museum of Northern Arizona; 1978:13–24.

Diamond, J. The golden age that never was. Discover 9(12):70–79; 1989.

Dobyns, H. F. From fire to flood: Historic human destruction of Sonoran Desert riverine oases. Soccoro, New Mexico: Ballena Press; 1981.

———. Their number become thinned: Native American population dynamics in eastern North America. Knoxville: University of Tennessee Press; 1983.

Fagan, B. M. People of the earth: An introduction to world prehistory. Boston: Little, Brown and Company; 1986.

Feely, J. M. Did Iron Age man have a role in the history of Zululand's wilderness landscapes? South African Journal of Science 76:150–152; 1980.

Frison, G. C. Prehistoric hunters of the high plains. New York: Academic Press; 1978.

Grayson, D. K. Death by natural causes. Natural History 96(5):8–11; 1987.

Guthrie, D. Primitive man's relationship to nature. Bioscience 21:721–723; 1971.

Hill, K.; Hawkes, K. Neotropical hunting among the Ache' of Eastern Paraguay. In: Hames, R.; Vickers, W., eds. Adaptive responses of native Amazonians. New York: Academic Press; 1983:139–188.

Hill, K.; Kaplan, H.; Hawkes, K.; Hurtado, A. M. Men's time allocation to subsistence work among the Ache' of Eastern Paraguay. Human Ecology 13:29–47; 1985.

Jacobs, W. The white man's frontier in American history: The impact upon the land and the Indian. In: Jacob, W., ed. Dispossessing the American Indian. New York: Charles Scribner's and Sons; 1972:19–30.

Jennings, J. D.; Schroedl, A. R.; Holmer, R. N. Sudden Shelter. University of Utah Anthropological Papers Number 103. Salt Lake City: University of Utah Press; 1980.

Kay, C. E. Too many elk in Yellowstone? Western Wildlands 13 (3):39–44; 1987. Review of: Despain, D.; Houston, D.; Meagher, M.; Schullery, P. Wildlife in transition: Man and nature on Yellowstone's Northern range. Boulder, CO: Roberts Rinehart, Inc.; 1987.

———. Yellowstone's northern elk herd: A critical evaluation of the "Natural Regulation" paradigm. Logan: Utah State University; 1990. Dissertation.

Kelly, R. L.; Todd, L. C. Coming into the country: Early Paleoindian hunting and mobility. American Antiquity 53:231–244; 1988.

Layng, A. American Indians: Adapting to change. USA Today Magazine (Society for the Advancement of Education). 1986. Sept.

Lewis, H. T. Patterns of Indian burning in California: Ecology and ethnohistory. Ballena Press Anthropological Papers No. 1. Ramona, CA: Ballena Press; 1973.

Lister, R. H. Contributions to Mesa Verde archaeology III: Site 866 and the cultural sequence at four villages in the Far View group, Mesa Verde National Park, Colorado. Boulder: University of Colorado Studies, Series in Anthropology 12; 1966.

Lucius, W. A.; Colville, J. K. Osteological analysis of faunal remains. In: Jennings, J. D.; Schroedl, A. R.; Holmer, R. N., eds. University of Utah Anthropological Papers Number 103. Salt Lake City: University of Utah Press; 1980:157–169.

McCleod, W. C. Conservation among primitive hunting peoples. Scientific Monthly 43:562–566; 1936.

Martin, C. L. Fire and forest structure in the Aboriginal eastern forest. Indian Historian 6(3):23–26; (4):38–42; 1973.

———. The American Indian as miscast ecologist. The History Teacher 14:243–252; 1981.

Martin, P. S. Pleistocene overkill. Natural History 76(10):32–38; 1967.

Matson, R. G.; Lipe, W. D. Settlement patterns on Cedar Mesa: Boom and bust on the northern periphery. In: Euler, R. C.; Gumerman, G. J., eds. Investigations of the Southwestern Anthropological Research Group. Flagstaff: Museum of Northern Arizona; 1978:1–12.

Meade, J. I.; Meltzer, D. J. Environments and extinctions: Man in late glacial North America. Orono: Center for Study of Early Man, University of Maine Press; 1985.

Nabokov, P., editor. Native American testimony. New York: Harper and Row; 1978.

Nichols, R. F.; Smith, D. G. Evidence of prehistoric cultivation of Douglas-fir at Mesa Verde. In: Osborne, D., editor. Contributions of the Wetherill Mesa Project. Washington, DC: Society for American Archaeology Memoir 19; 1965:57–64.

Nudds, Thomas D. The prudent predator: Applying ecology and anthropology to renewable resources management. In: Novak, M.; Baker, J. A.; Obbard, M. E.; Malloch, B., eds. Wild furbearer management and conservation in North America. Ontario, Canada: Ministry of Natural Resources; 1987:113–118.

O'Connell, J. F.; Jones, K. T.; Simms, S. R. Some thoughts on prehistoric archaeology in the Great Basin. In: Madsen, D. B.; O'Connell, J. F., eds. Man and environment in the Great Basin. Society for American Archaeology Papers 2. Washington, DC: Society for American Archaeology; 1982:227–240.

Pringle, H. Habitat: A plaster disaster. Equinox 9(6):125; 1990.

Pyne, S. J. Fire in America: A cultural history of wildland and rural fire. Princeton, NJ: Princeton University Press; 1982.

Ramenofsky, A. F. Vectors of death: The archaeology of European contact. Albuquerque: University of New Mexico Press; 1987.

Schlanger, S. H. Patterns of population movement and long-term population growth in southwestern Colorado. American Antiquity 53:117–126; 1988.

Simms, S. R. Acquisition cost and nutritional data on Great Basin resources. Journal of California and Great Basin Anthropology 7:117–126; 1985.

————. Behavioral ecology and hunter-gatherer foraging: An example from the Great Basin. Oxford, England: BAR International Series 381; 1987.

Smith, E. A.; Winterhalder, B. New perspectives on hunter-gatherer socio-ecology. In: Winterhalder B.; Smith, E. A., eds. Hunter-gatherer foraging strategies: Ethnographic and archeological analyses. Chicago: University of Chicago Press; 1981:1–12.

Speck, F. G. Aboriginal conservators. Bird-Lore 40:258–261; 1939.

Speth, J. D.; Spielman, K. A. Energy source, protein metabolism, and hunter-gatherer subsistence strategies. Journal of Anthropological Archaeology 2:1–31; 1983.

Steward, J. H. Basin-Plateau aboriginal sociopolitical groups. Washington, DC: Bureau of American Ethnology Bulletin 120; 1938.

————. Culture element distributions XIII: Nevada Shoshoni. University of California Anthropological Records, Berkeley 4(2):209–360; 1941.

————. Culture element distributions XXIII, Northern and Goshiute Shoshoni. University of California Anthropological Records, Berkeley 8(3): 263–292; 1943.

Stewart, O. C. Culture element distributions XIV, Northern Paiute. University of California Anthropological Records, Berkeley 4(3):361–446; 1941.

————. Culture element distributions XVIII, Ute-Southern Paiute. University of California Anthropological Records, Berkeley 6:231–355; 1942.

Stiger, M. A. Mesa Verde subsistence patterns from basketmaker to Pueblo III. The Kiva 44:133–145; 1979.

Strickland, R. The idea of the environment and the ideal of the Indian. Journal of American Indian Education 10:8–15; 1970.

Thomas, D. H.; Mayer, D. Behavioral faunal analysis of selected horizons. In: Thomas, D. H., ed. The archaeology of Monitor Valley 2. Gatecliff Shelter. Anthropological Papers of the American Museum of Natural History 59(1); 1983:353–391.

Thornton, R. American Indian holocaust and survival: A population history since 1492. Norman: University of Oklahoma Press; 1987.

Todd, L. C. Analysis of kill butchery bonebeds and interpretations of Paleo-indian hunting. In: Nitecki, M., ed. The evolution of human hunting. New York: Plenum Press; 1987:225–266.

Wheat, J. B. The Olsen-Chubbuck site: A Paleo-Indian bison kill. Society for American Archaeology Memoir Number 26. Washington, DC: Society for American Archaeology; 1972.

White, R. Land use, environment and social change: The shaping of Island County, Washington. Seattle: University of Washington Press; 1980.

————. Native Americans and the environment. In: Swagerty, W. R., ed.

Scholars and the Indian experience: Critical reviews of recent writing in the social sciences. Bloomington: Indiana University Press; 1984:179–204.

Winter, J. C. Hovenweep 1976. Archeological Report No. 3. San Jose, CA: San Jose State University; 1977.

New

Management

Concepts

Ecological Education
A Contemporary Imperative

CHERYL CHARLES

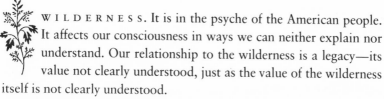 W I L D E R N E S S. It is in the psyche of the American people. It affects our consciousness in ways we can neither explain nor understand. Our relationship to the wilderness is a legacy—its value not clearly understood, just as the value of the wilderness itself is not clearly understood.

The American psyche contains contradictions related to wilderness. We have, historically, wanted to chart unnavigated waters and map un-walked forests and canyons; we have evidenced a need to explore the unknown and master it in some sense. Simultaneously, we yearn to keep the wilderness, wishing it were not all mapped, wishing there were still new frontiers. We have feared the wilderness, seeking to conquer it; we have cherished the wilderness, wishing to protect it in its wild state.

To talk about wilderness in the American psyche then is to talk about much more than simply a place for wildlife to live in a natural state, to talk about much more than the need to protect gene pools as precious repositories of biological diversity—as important as both of those things are. Wilderness provides an indefinable nutrient within the whole of the American body—a body of intellectual, spiritual, emotional, and physical character. Wilderness is an important nutrient for the whole of the body of planet Earth.

I keep referring to the *American* psyche—thinking of North America, and especially Canada and the United States. That is only because I know more about the North American experience. I am not intentionally excluding the psyche of the planet's other people. I suspect similar qualities to be inherent in all humanity. For some cultures, the direct link to the wilderness in daily living is simply in the more distant past than it is in the United States and Canada in particular.

An anecdote illustrates the likelihood of a shared, intrinsic yearning for wilderness among all the peoples of the world. Last year, while in India working as a consultant to the National Centre for Environment Education, in Ahmedabad, I was asked to assist in developing a national strategy for environmental education. The Centre would like to be the national sponsor for Project WILD in India, and is at least as interested in our approaches to the processes of leadership development, volunteerism, networking, and teacher education as in our Project WILD materials. Early in my time there, it seemed important to pose the question, "Why care? Why care about wildlife and wild lands in a nation of nearly one billion people, almost twenty percent of the entire planet's population, where clean water, food, suitable shelter, and space in which to live are not easily available for literally millions of the nation's inhabitants?" So I did ask the question, "Why care?" What response did I get? People answered, with elegant conviction—as if there were no question—"It is not enough for humans only to survive. It is not enough."

It is not enough, anywhere, for humans only to survive—first on ethical, and then on psychological, aesthetic, scientific, ecological, economic, and recreational grounds. From grounds that involve human use to those that involve humans not at all, there is value to wildness.

To say that there is value to wildness and wilderness does not say that there is clear agreement and understanding within our society about the value of wilderness. In fact, that too is the wilderness legacy—the confusion between the manifest destiny to conquer and the intrinsic value of wildness in its natural state.

Even if all of society today agreed on the importance of wilderness, we are rapidly losing within the population a clear and accurate understanding of the nature of wilderness and the character of its many values. So, even among those who say that wilderness is important only for its own sake, we have a society that increasingly feels alienated from natural systems and lacks an understanding of how wilderness works and what it needs to survive.

As individuals within postindustrial societies such as the United States, we have grown less responsible for securing our own food and meeting our other physical needs. We have less opportunity to experience directly how the world works. As a result, when we do go outside, we inadvertently cause damage to wild animals, to plants, and to whole environments, now, and consequently for future generations to an increasing and alarming degree. These actions affect all living environments; even set-aside wilderness areas are not immune to human degradation. In domesticated environments, for example, we plant lawns in areas with limited water supplies. We introduce exotic wildlife and vegetation that jeopardize native plants and animals. We drain wetlands and spray roadside vegetation, destroying natural habitats for wildlife. We regularly eliminate whole habitats, destroying them as places in which wildlife can live. We scar the earth with our recreational vehicles, damaging vegetation, creating erosion which affects water quality, and leaving marks upon the earth that will be there when our grandchildren have grandchildren. We dump pesticides in our water supplies. We simply do not know how to take care of the garbage we produce and the toxins we generate. We poison ourselves, our children, the plants, animals, skies, waters, and the living earth that is our home.

In the face of all of this, our society needs education for ecological literacy. It is a contemporary imperative. In fact, education for ecological literacy must be the foundation for all education in the twenty-first century. However, education about the environment is not a priority in most of the kindergarten-through-high-school classrooms of North America. That is why Project WILD's work is so important.

Project WILD is a conservation and environmental education program especially designed for teachers of kindergarten-through-high-school-age youth. Through instructional workshops for teachers as well as the materials and continuing services we provide, we prepare teachers to integrate concepts about wildlife, the environment, and responsible human actions into all major subject and skill areas. Since most of today's schools do not identify education about the environment as a priority, we work to make responsible and effective matches with what *are* the educational priorities in the schools.

Basic skills of reading, writing, and arithmetic, critical thinking, and decision making form the dominant curricular emphases in elementary schools. These are obviously appropriate and important emphases; however, they do not need to be taught by *excluding* the foundation of education about the environment. Even science education is limited in its scope and effectiveness. For example, there isn't much life in life science, especially in the high schools. Yes, there has been an emphasis on excellence in science education, but, with the exception of high school science instructors, most teachers are reluctant to teach anything that sounds like science. Education about wildlife and the environment, including wilderness, sounds like science. Although there are a growing number of exceptions in public and private schooling, in the mainstream of formal education in North America, this description fits the norm. Teachers tend to lack confidence, knowledge, preparation, and instructional materials to help them teach about the value of a healthy environment, including the importance of wilderness. So we with Project WILD are working to help with the confidence, knowledge, preparation, and materials that tend to be lacking.

Although the concept of Science, Technology, and Society (STS) has been popularized in recent years as important to the curriculum, STS typically considers science-technology-and-society separate from the larger context of the living environment. Dr. John Disinger of Ohio State University has been among those championing the concept of STSE—Science, Technology, Society, and Environment. My husband, Bob Samples, and I have taken a holistic approach to STS issues for years, editing a book on the subject in 1978, long before it had become a buzz phrase in education. Most teachers, however, if they include STS at all, do not emphasize an ecological perspective in the process.

At the high school level, most teachers use a textbook-and-lecture approach most of the time, with perhaps as many as ninety percent of the teachers using this approach ninety percent of the time. At both the elementary and secondary levels, most teachers do not take their students outside to study and learn in the first classroom, the living world itself.

Add to this brief characterization of formal schooling in North America some of the additional external influences of our society. For example, consider the pressures of changing family models, economic and social hardships, the pervading dangers of drugs and their availability to young people, and the need for AIDS education. Readily apparent is the reason so much of the education establishment focuses on youth at risk. All of this points to the reason education about wildlife, wild lands, and the environment as a whole is not a visible priority in most schools in North America.

The education establishment tends to follow, not lead. That is not a criticism. That is just the nature of public schooling in our society. Schooling as a whole is designed to transmit the values of the culture, not to establish those values. As a result, the education establishment responds to society's leadership. Society's leadership in turn is dependent upon the leadership of individuals. That should suggest, in the least, that we have a responsibility to exercise leadership. We also have a responsibility, I believe, to ask for leadership—for example, from our elected officials.

Be encouraged. People throughout society are saying, in increasing numbers, that education is the key to sustaining the quality of the environment.

I define education as a process by which a learner is enabled to do something. Conservation and environmental education focus in a way that enables a learner to take responsible action on behalf of a sustainable future. We have outcomes in mind. We want learners to take responsible action to benefit the health and quality of the environment, now and in the future.

By definition, the process of education requires that learners acquire awareness, knowledge, attitudes, skills, experiences, and commitment to result in informed decisions, responsible behavior, and constructive actions. People must be engaged in a process of education that includes access to accurate information to develop a base of knowledge. But information alone, even when accurately and effectively presented, is not enough. In addition, they need to develop their awareness, to understand "Why care?" They need to develop attitudes and an ethic that will guide them through life's choices. They need a variety of skills, including those for decision making in the face of complex issues. They need opportunities to develop and apply their knowledge. *They need opportunities to succeed in making a difference.* When learners perceive they have skills that can enable them to accomplish something of value to them, that is when they make the shift from apathy to action. All of these things: awareness, knowledge, attitudes, skills, and experiences combine to create a commitment to live a life of responsible action that is dedicated to conserving, protecting, and enhancing the health and quality of the environment.

As it turns out, supporting and implementing this kind of well-conceived and thoughtful approach is recognized as good education. That is why, when leaders in the education establishment do learn about conservation and environmental education, they tend to find it appealing, valuable, and supportive of their instructional goals.

Enormous long-term payoffs accrue from preparing young people for ecological literacy throughout their lives. The process of education for ecological literacy must, however, be available to learners of all ages. For at least two reasons, we must also address ourselves to the adult public:

1. The ecological threats are so great that we need to create an informed decision-making public now. We can't just wait for the next generation to correct all of our excesses and errors.

2. There is an escalating cry for ecological education from the general public. The recognition by *Time* magazine in selecting the Earth as Planet of the Year, rather than selecting typically a "Man of the Year," represents a genuine, deeply held, and growing concern within the populace for the serious ecological conditions facing the whole of the planet and its inhabitants. Pollsters, like Harris and Roper, for years have been finding such a persistent concern for environmental quality that they describe it as a trend, not a fad. People want to know what to do. They are concerned. They want accurate information. They want assistance. They want leadership. A comprehensive approach to ecological education seems to be the key.

We do have to translate our concerns into committed action. Opportunities and responsibilities exist in the formal arenas of elementary, secondary, and university curricula to create an ecologically literate public. They also exist in a variety of other ways—through the media; in recreational settings; in the business community; through civic groups; in churches; in parks, aquaria, and zoos; and, obviously, through private conservation and environmental organizations. Baby-boomers and long-lived people form two enormously important and influential segments of this society that we must engage in a commitment to education for ecological literacy.

So what must each of us do to help create an ecologically literate society? Here are some suggestions.

1. *Become an advocate for ecological literacy.* Without taking an alarmist stance, speak clearly and authentically about the responsibilities we have to conserve, protect, and enhance the quality of the environment, now and for future generations. Without an ecologically literate public, we will be ill prepared to solve the ecological crises we now face or to prevent their escalation into widespread changes affecting continued life on this planet.

2. *Encourage school officials and teachers to integrate education about the environment into the fabric of kindergarten-through-high-school curricula.* Environmental education—personal, societal, and ecological—provides a natural organization for school learning. It helps students see relevance in their schooling. It provides a foundation and a context within which to use awareness, knowledge, attitudes, skills, and experiences to live lives of efficacy and well-being.

3. *Champion the idea that it is good business to have a healthy environment.* For the long term, it is not profitable to have a degraded environment. Short-term payoffs may be large in a monetary sense. Even monetarily, in the long term, ecological resources have to be abundant and healthy for economic profitability to be sustained. Increasingly, leaders in the business community of this society are standing up, stepping forward, and taking a stand for protecting natural resources. These business leaders should be acknowledged, commended, and encouraged. At a local level, perhaps you are in business or have friends who are. Encourage business people to identify the natural resources that their business relies upon, either directly or indirectly. The first important step is often simply to recognize the resource base that is so often taken for granted. What is the source of the electricity used in the business? Is either recycled or recyclable paper being used? Is waste paper being recycled? Even in high-tech industries, the reliance on natural resources is enormous. People often simply need a little encouragement and assistance in recognizing this ecological foundation and, then, in conserving the resources involved.

4. *Create opportunities for people to get outside, into the first classroom, the natural world.* Everyone does not need to hike and camp in the wilderness. Although that experience is personally valuable for

nearly everyone, the potentially huge numbers of people who could take advantage of that opportunity today on this planet will in themselves have a degrading impact on the wilderness itself without some limits and management of their access. However, the fact is that most Americans will not take a ten-day wilderness excursion each year. But most people will at least step outside. More and more, we need to create opportunities for people to learn and enjoy the out-of-doors, intentionally grounding their appreciation for the importance of natural outdoor environments in the process.

5. *Teach people the values of wilderness.* Wilderness is important —for tangible and intangible reasons. Given the confused history of humanity's relationship to wilderness areas in North America and the world, we must take the initiative to teach people about these values. As just one example, since education about the environment is not a visible priority in most schools in North America, education about the importance of wilderness certainly is not either. We need to teach the values of wilderness.

6. *Avoid zealotry.* Zealotry dissuades more than persuades. It creates antagonists more than partners. We need to create a climate of shared responsibility for the quality of life on this planet. Cooperation and education are allies in this process. We don't have time for a reactionary backlash. We need to go forward in good spirit, with good sense and good works, to create and protect healthy environments with every individual citizen feeling empowered and enfranchised to assist in the process.

7. *Nourish yourself.* Take time for yourself, alone, with friends and loved ones, to get outside, and into the wilderness whenever you can. It is an important source of personal inspiration and sustenance to spend time outside with the source. Do that, for yourself, for the conviction those experiences will continue to bring to your words to others, and, in the long term, for the wilderness itself. Without those who continue to know it well, who love it, and who will speak on its behalf, wilderness itself is threatened.

The time is right, our efforts are coalescing, tangible progress is being made. We need to increase our efforts, pay attention to quality, work in cooperation with our intended audiences, and exercise a tremendous amount of initiative and creativity while grounding ourselves in good science. We live in the midst of a throwaway society, and we are part of a culture that acts as if this were a throwaway planet. We know it is not. We must each help to bring that message home. Earth is home to us all. Water is life to us all. Wildness and wilderness are necessary for the life of the human spirit, as for the health of the body of the planet itself. Education to appreciate, conserve, protect, and enhance life on this earth is essential to our present and vital to our future.

Environmental Economics and
Wilderness Values
What's In It for the Individual?

RICHARD M. ALSTON

CLIFFORD R. NOWELL

JANUARY 1989 marked the first issue of *Ecological Economics, The Journal of the International Society for Ecological Economics. Ecological Economics* hopes to become the vehicle for extending and integrating the fields of ecology and economics. Echoing the theme of the First North American Interdisciplinary Wilderness Conference, the editorial board argues that such an integration is necessary because conceptual and professional isolation have led to economic and environmental policies which are mutually destructive rather than reinforcing. The changing nature of the debates about wilderness and environmental policy indicates that the emergence of such an interdisciplinary journal has been long awaited by those who struggle with the basic economic and ecological paradigms and their underlying assumptions.

The emergence of an interdisciplinary concern for environmental management was heralded twenty-five years ago by the passage of the Wilderness Act of 1964. There soon followed the National Environmental Policy Act (NEPA) of 1969, the Clear Air and Water Act, and numerous subsequent pieces of legislation aimed at insuring judicious use, protection, and stewardship of wild lands and natural resources in the United States.

Over the years and with few exceptions, the participants in the inevitable and ongoing debates have not agreed on specific solutions, but they have shared the desire to find methods for implementing efficient environmental policies. Those who dedicated themselves to the principles of integrated and interdisciplinary resource management have long since found that, at bottom, environmental issues almost always come down to a basic question: "Is this or that action, this or that policy, this or that wilderness designation worth the cost?" Environmental and ecological issues, in other words, are simultaneously economic issues as well. One side of the debate has defined the issue as follows:

> When the Forest Service pursues management of areas where returns do not cover costs, it diverts scarce investment capital from more productive lands where the public money would be better spent. In addition, the subsidizing of federal timber buyers undermines investment on private lands, where industrial and other owners must recover all management cost to make a profit, yet must compete with the federal timber in the marketplace. (Sample and Kirby 1983, 53–54.)

All that jargon—returns, costs, investment capital, productive lands, money, buyers, profit, competition—sounds just like an economist, doesn't it? But it is not an economist speaking. Indeed, the foregoing quotation is an expression of conservationist concern in a conservationist's handbook published collectively by The Wilderness Society, Sierra Club, National Audubon Society, Natural Resources Defense Council, and the National Wildlife Federation (Sample and Kirby 1983). The statement of concern continues by arguing that "conservationists believe there should be rigorous economic analysis" in such areas as determining timberland suitability, trade-offs between developed and dispersed recreation, and other aspects of wildland management. Peter Emerson, an economist working for The Wilderness Society has stated that "the environmental community is striving to put . . . the future of our national forests into the larger economic and social context. . . . We believe that our efforts are in tune with economic reality." Siding with mainstream economic philosophy, he suggests that environmentalists should use the present value (efficiency) criterion long advocated by economists to evaluate development projects because "it is the standard analytical procedure used to measure net returns to projects involving receipts and expenditures at different points in time" (Emerson 1987, 1–7).[1]

More recently, environmentalists and wilderness advocates have used economic analysis as a critical, if not central, element in conducting litigation to stop what they view as the needless rape of the public lands by agencies such as the Forest Service and the Bureau of Land Management. According to the environmentalists, economic decisions taken by these agencies have resulted in "money losing timber sales [that] are costing taxpayers at least $250 to $500 million dollars per year" (O'Toole 1988, xi). The Forest Service, they argue, often subsidizes logging and logging jobs at a cost far in excess of what the jobs or the timber are worth. Based on what environmentalists consider unwarranted estimates of rapid and continuing growth in developed recreation demand, the agencies construct permanent roads into previously virgin, roadless, wilderness-candidate areas. And, as if to add strength to their case, the environmentalists add that this apparently wanton destruction of the nation's inheritance is being done at the taxpayers expense. "Why," the environmentalists ask rhetorically, "do the professional public land managers destroy these areas if they continuously lose money?" (O'Toole 1988, 2).[2]

If, as many environmentalists now claim, the answer lies in economics and economic incentives, then it is time for those who want to maximize wilderness objectives to come to grips with the fact that environmental economics and wilderness values are inseparable. Indeed, the wise use and protection of the existing and yet to be dedicated wilderness land in America most likely depends on the careful application of ecologically sound economic analysis by fully informed citizens and environmental advocacy groups.

In this essay, we describe the analytical tools of environmental economics and explain how they may be applied to improving wilderness policy. We focus on the economic approach to the benefits and inevitable costs that accompany society's choice to have more or less environmental amenities, such as wilderness. We make explicit the message that these values can and ought to be measured. But we also make explicit the underlying epistemological assumption behind virtually all economic analysis—the central place held by individuals in valuing resource management outcomes. The purpose is to arm the reader with an awareness of both the strengths and weaknesses of economic analysis in wilderness policy debates.

The Economic Way of Thinking

Economics is the study of how society chooses to use its limited resources. A relatively new, but increasingly important branch of economics is environmental economics. The roots of environmental economics are planted firmly in mainstream economic theory. Practitioners apply the insights of economic theory to the study of how we choose to allocate scarce resources such as endangered species, scenic vistas, water, and air among all possible competing human ends.

Underlying all economic analysis is the assumption that environmental amenities may be studied and understood in the same manner as any other economic good. Based on that assumption, society—in its attempt to satisfy the myriad demands for other goods and services—may clearly end up with too little of the environmental amenities it simultaneously desires. When that is the case, processes are initiated that attempt to correct the situation. More and more resources are poured into planning, legislating, and litigating the matter, all in the attempt to increase the output of amenities available from environmental resources and wilderness areas.

Just as economic theory recognizes that it is possible to have too few environmental amenities, the economic way of thinking also suggests that it is possible to have too many (i.e., too much environmental quality, too many bald eagles, too many red-cockaded woodpeckers, too many northern spotted owls, too many grizzlies and eastern grey wolves, and too much clean air). Although the idea of an air or water quality which is too great may seem impossible to some, environmental economics forces us to recognize that increasing the level of environmental quality requires forgoing other valuable alternatives. Environmental quality is not available without cost.

Since there are alternative uses of our natural resources, we must choose among them. The ends we serve and, in some instances, the gains to be had by expanding the amount of a particular environmental amenity may force us to forgo extremely valuable alternatives which the majority of society's members do not deem to be worth the cost. But the aforementioned insights are little more than common sense. Enter ecologists and economists!

Ecologists can attempt to tell us the conditions under which northern spotted owl and grizzly bear populations prosper. They can estimate how many acres and what types of habitat the owls and bears use when unhindered by human activity. Ecologists may even predict herd losses in response to rising grizzly populations. But they cannot tell us how many mating owl pairs or grizzly dens is optimum (or optimum for what?). Economists can explain the trade-offs involved between unlogged, roadless old growth forests and jobs in the mill town. They can tell us the implicit values behind a choice to operate a chemical plant rather than protect riparian habitat and a valuable fishery. But they cannot tell us the proper level of either. Indeed, neither economists nor ecologists have any special expertise that suggests that they should, let alone could, tell society how much pollution should be tolerated, how much of the country's backcountry should be designated wilderness, or how many acres of wetlands should be set aside as a preserve for migrating wildfowl. Ecologists can predict the impacts. Economists can describe the choices. But neither has a pipeline to the Holy Grail. Equally important, neither economists nor ecologists can provide conclusive information about any of the things mentioned herein. If they could, there would be no need for numerous ongoing, multidisciplinary, multiagency research projects which have yet to resolve problems regarding many environmental issues.[3]

Rather, economics and ecology only provide tools to help society determine the impacts and associated costs and benefits of achieving a certain level of environmental quality. An overview of the techniques used by economists to evaluate the costs and benefits of environmental choices follows, as does a discussion about how such information might be used to help society choose the desired levels of environmental quality, resource outputs, and, in general, alternative uses of our rich natural endowment.

Alternatives, Choices, and Opportunity Costs

To determine the costs of producing any good (including environmental quality), economists rely on the concept of opportunity cost. The opportunity cost of any action is the loss of the next best alternative to that action. Thus, the cost of clean air is not the dollar amount spent removing the pollutants, but rather it is the cost of what might have been achieved had the resources (represented by the dollars necessary to command them) been used elsewhere. Similarly, the cost of building roads, logging, and developing a previously roadless primitive area is the lost opportunity for dispersed recreation (including the decrease in the value caused by fewer wildlife encounters for sightseers and hunters and by spoiled scenic vistas). Similarly, the opportunity cost of *not* building roads and logging a pristine mountain setting must include the costs both of increased lumber prices and the possibility that because of the price increases, people choose to use wood substitutes such as plastics which carry their own environmental costs. Opportunity cost measures the value of the chosen alternative in terms of the sacrificed or foregone next best alternative. The tradeoff between the use of wood and plastics suggests the complications in defining the next best alternative.

Economists often illustrate the concept of opportunity cost by defining the set of all possible combinations of goods and services that could be produced when the economy is using its resources to full potential. Imagine, if you will, an economy that produces only two types of outputs: on the one hand, environmental goods (e.g., scenic vistas, clean air, dispersed recreation in wilderness areas) and, on the other hand, all other things that people desire.

Think of this set of potential combinations as representing the menu of choices which confront society. The members of society could, of course, choose to devote *all* the available resources to achieving the maximum possible level of environmental quality. If it made that choice, however, the society would have no unused resources with which to produce other things. Alternatively, they could ignore the impact on environmental quality and use all of the available resources to produce other things. Without exception, of course, societies choose neither extreme. They choose a bundle of outputs that contains some of both of the outputs (goods) they desire.

It is important to clearly understand the concept of "full potential." If society wastes resources, uses them inefficiently, or fails to allocate resources where they are the most productive (i.e., at their highest and best use in each situation), then clearly society has forced itself off the menu. The choices available off the menu include all those combinations which could be improved upon simply by using existing resources more efficiently and in accordance with their relative productivity (economists would say according to their comparative advantage). It makes no sense, for example, to harvest timber using costly aerial and helicopter logging techniques on low-grade, highly erodible, forest land near riparian, fish-spawning habitat, while highly productive, forest land on stable soils far from spawning habitat is dedicated to lower valued uses. By reallocating its resources, the members of society could have more environmental quality without giving up anything else. Alternately, it could have more of the other things that it wants without having to sacrifice environmental quality in the process. Or it could have more of both. The environmental community needs no convincing on this matter, as evidenced by the recently issued *Blueprint for the Environment* (Maize 1988).[4]

> The new administration has an extraordinary opportunity—to both strengthen the economy and improve the environment via least-cost energy policy. Such a policy would ensure that government programs and spending are allocated so that the most cost-effective means to supply energy services and reduce energy-related pollution are pursued first. . . . By avoiding wasteful projects like synthetic fuels development and nuclear breeder reactors, the federal energy budget can actually be reduced while the national energy picture is substantially improved. (12–13)

Whether you agree with this statement, the point, of course, is that everyone's interests may be served when resources are used efficiently.[5]

The issue, then, is not whether we should use resources to their full potential in producing the things that we want. Rather, the issue is how we are to choose among the infinite possible combinations that face us when resources *are* allocated efficiently. Which of an almost infinite number of efficient alternative bundles of environmental amenities and other goods that are available on the menu of choice should be chosen? What specific combination of environmental quality and other goods will society choose? That depends on the leaders we elect, the institutions we create, and the preferences of people expressed through the ballot box and the marketplace, as well as through arbitration, litigation, and political compromise.

And what of the nature of that choice? Ecologists often envision the set of choices as if it were a plateau. Anywhere on the plateau is simply a matter of choice. But there are thresholds or safe sets beyond which we move at our own peril. If we are unwittingly near the edge and choose to have still more goods, we are likely to plummet off the cliffs into an unknown abyss. Economists, on the other hand, think of the choice set as if it were a geometrically increasing curve. The ever rising curve demonstrates the fact that the opportunity cost of the first few units of consumer goods may be achieved at little cost in terms of environmental quality. But as more and more resources are devoted to obtaining even more consumer goods, the opportunity cost in terms of environmental destruction rapidly increases. Beyond some point, the extra consumer goods are not worth the opportunity cost. Of course, the trade-offs cut both ways.

The Pollution Standards Index (PSI), for example, converts levels of air pollution to a scale that runs from 0 to 500. When the index takes on values of 100 or more, a possible health risk exists and the air quality is classified as unhealthful; a level of 300 is considered hazardous. Reducing the level of the index from 500 to 300 by removing large amounts of pollution in a manufacturing process that burns coal is relatively simple and can be achieved by installing low-cost scrubbers to reduce emissions. We observe reduction in the pollution levels with only a moderate opportunity cost. Once the pollution level reaches the neighborhood of 150, however, the cost of removing emissions requires that substantially more of the remaining resources be dedicated to the task. If the most productive resources of other goods have been saved until last (i.e., kept in the production of consumer goods), society will have to make large sacrifices of consumer goods to reduce pollution by a small amount. While we may not agree with the choice, many in society (including those who stand to lose jobs and income) deem the cost too high. They prefer a bundle on the menu of choice that contains less environmental quality (i.e., more pollution) and more of the other things they desire. The extra clean air isn't worth it.

Thus, a greater level of environmental quality and more areas dedicated to wilderness values would always be desired if they were free. But they are not. The fundamental economic problem, then, is the problem of choice. We must determine to what extent we can satisfy our goals within the constraint of limited resources.

The Supply and Demand for Economic Goods

Ecologists often speak of the delicate balances that exist in nature. Through an intricate web that is better described than understood, nature seems to allow species populations to ebb and flow in a series of interconnected, synergistic systems in equilibrium. In the absence of human intervention, nature seems to find a way over time to provide just the right amount of each and every species in the food chain. Disequilibria, of course, occur. But the equilibrating mechanism of nature seems to right the system over time. (Nature, of course, doesn't seem to care which species prosper, which decline, or which die off. That is a judgment placed on the outcome by only one species, *Homo sapiens*.)

In a similar way, economists identify the market system as a self-adjusting, integrated set of resource-allocation systems that tend to provide (over time) the right amount of goods and services. When, for example, was the last time that you went to purchase milk only to find there wasn't milk to be had? Of course, disequilibria occur. Often we are aware of temporary shortages of specific goods, but the problem does not appear to be chronic. A freeze in Florida may cause a shortage of orange juice, but in a few months or a year, the problem seems to have disappeared. Fluctuating prices provide the necessary signal to right the system.

If, as economists and a rising number of environmental groups argue, environmental quality is like any other economic good, then why do we tend to get too little of it? Why does the market supply society with an unacceptable level of pollution? Society has been plagued with a pollution problem for decades, yet this shortage of environmental quality does not seem to dissipate. The answer may lie in general market failure.

Market Failures

Consider the case of an electric utility which generates power by using coal. Competitive markets (or regulatory authorities) insure that the firm will be able to sell the electricity for a price to consumers that just covers the cost of production (i.e., the market value of the alternative uses for labor, natural resources, and capital). Not included in its recoverable costs, however, are resources that are used but that are not paid for through established markets. One such resource is air. A by-product of the generation of electricity is air pollution and acid rain. Society experiences an additional cost above what the utility must pay to produce energy. The social cost of polluted air is real and is borne by the society at large. But, unless some corrective action is taken, it is not included in the price paid by consumers of electricity. The private market has failed to take into account all of the costs of production. Economists refer to these uncounted costs as "externalities." Opportunity costs are understated.

By ignoring the social costs (as opposed to the market costs) of production, electricity is oversupplied while environmental quality is undersupplied. A popular view among economists is that the problem lies in the fact that no one owns the air. Ownership of other factors of production are well defined, and we can generally count on the owners to allocate their resources efficiently (i.e., to their highest and best use— as indicated by the price signals of the market). Opportunity costs involved in the use of such resources are thus taken into account. Not so with resources which we all own collectively.

If everyone has an equal claim on the resource, but no one has the ability to charge for its alternative uses, then the outcome is as if no one owned it. Economists call this the "tragedy of the commons." [6] Air, fishing grounds, and whales are treated as free goods. As a consequence, people do not choose carefully, and these resources are allocated to any and all possible uses, including those least valued. And why not? If the resource is indeed free to the user, then there is no opportunity cost involved in its use. Like the electric company, the individual user need not forgo anything. But, the cost doesn't disappear. It is borne by society at large.

Guiding Nonmarket Solutions with Economic Reasoning

If the market cannot supply society with the desired amount of pollution, wilderness, and other environmental goods because of the absence of property rights, what can be done? Can the government or some central authority step in and proclaim that the air and water resources are no longer free goods? In theory, the answer to this last question is clearly "yes." The government can define who owns the resources used in production, and the owner(s) can charge individuals for polluting the environment, for using the wilderness, and so on.[7]

With or without clearly defined property rights, however, society constantly faces the choice of increasing or decreasing the amount of resources devoted to environmental quality. Decisions made at the margin (i.e., when the consideration is between "a little more of this, a little less of that") require that somehow we collectively consider the extra costs and benefits of an increase or decrease of any activity that impacts the environment. If the marginal costs outweight the marginal benefits, the activity should not be undertaken. Alternatively, if the additional (or marginal) benefits of an activity are greater than the additional (or marginal) costs, the activity should be undertaken.

Economists argue that it is possible to define and quantify most of the marginal benefits and marginal costs of every environmental issue in a manner that provides information essentially similar to that provided automatically by the market. With that information they can determine the level of pollution that people want and are willing to pay for (i.e., are willing to tolerate given the other options before them). Accepting for the moment that the economists can fulfill their promise, how does that information make it possible to achieve the "correct" level of environmental quality? We present three possible alternatives: applying direct quantity controls, collecting taxes and paying subsidies, and creating property rights in the environmental good where none previously existed.

QUANTITY CONTROLS

Many people advocate the imposition of standards (e.g., automobile emission standards) aimed at achieving the desired level of environmental quality. Although quantity controls are often used by federal, state, and local governments, the control is only as effective as the penalty which enforces the standard. What happens when the standard is violated? If the penalty is not sufficiently large and consistently enforced, violators will find it in their interest to evade the law. The standard will be meaningless (as anyone who has registered a car in a nonregulated county or removed a catalytic converter knows).

Quantity controls apply as well to the designation of wildlife refuges, wilderness areas, and parks. The fiat approach, however, often breaks down due to the extent of the areas normally considered for such designation and the vast numbers of conflicting interests involved. Because the political process is viable only through compromise, the "correct" quantity is virtually never achieved.

TAXATION AND SUBSIDIES

In the case of pollution control, the government could impose a per unit tax on environmental discharge equal to the calculated marginal social cost. Polluters would be forced to recognize the environmental consequences of their actions. Polluting firms and individuals would often find it less expensive to clean up the environment than pay the per unit tax on each unit of discharge. Even if the firms continued to use the atmosphere as a garbage dump because they find it cheaper to pay the tax than to alter their production process, the tax could (but need not necessarily) be used to clean up after the fact. The tax can be set at many different levels. The higher the per unit tax the lower the pollution levels. Subsidies to reward environmentally sound practices also provide an alternative. Both taxes and subsidies recognize the fundamental economic principle that people respond to incentives. If the incentives remain unchanged, so probably will the outcomes. Change the incentives, however, and over time even the existing decision makers are likely to alter their behavior.

As with a quantity control, the government must be able to identify the marginal benefit and marginal cost associated with pollution. Thus, the effectiveness (and political acceptability) of government control is predicated on an accurate estimate of marginal benefits and marginal costs. Of course, for many issues such as acid rain, global warming, and deforestation of the tropical forests, decisions must be taken before there is any hope of obtaining accurate estimates of the interactions involved, let alone obtaining marginal benefit and marginal cost information.

BARGAINING SOLUTIONS

Suppose the rights to pollute are sold at an auction. Potential bidders will include polluters as well as conservation groups who wish the air to remain clean. The bids made by the conservation groups will represent the value of clean air. They will not bid any amount greater than the clean air is worth, but they would be willing to pay their valuation of clean air. Whereas the conservation group will be willing to pay to insure that the air is not treated as a garbage dump, the polluter will be willing to pay for the ability to pollute the air. The bids will be a reflection of their respective valuation of the air.

The winner of the auction will be the polluter if the marginal benefits of the air to the polluter are greater than the marginal loss (or cost) of the air to the conservation group. Regardless of who wins the auction, the air is put to its most valued use. In this solution the government is not forced into trying to estimate the marginal cost and benefits from changing levels of environmental quality. The cost of implementing this program in a few cases may be relatively small compared to the other two proposals.

Two major problems accompany the bargaining solution. First, the transaction's cost may preclude any meaningful bids. That is, the cost of gathering together all interested parties may simply be prohibitive. Second, the bargaining solution invariably will be plagued by what economists term the "free rider problem." Clean air is a "public good," a technical term used by economists to denote situations where people cannot be excluded from enjoying the benefits. That is, the benefits and enjoyment of breathing clean air can be had by all regardless of whether or not he or she contributed to purchase the clean air. Because individuals cannot be excluded from the benefits of the clean environment, they do not have an incentive to pay for it. Public television is a classic example of the free rider problem. Many people watch public television but decline to contribute during the fund drive.

It has been suggested that in special cases government intervention is only necessary to define which party (the polluter or the conservation group) owns the property right to the use of the air (Stroup and Baden 1983). Consider the likely result if the conservation group is given the property rights to the air. An industry wanting to pollute the air could only do so if it were able to purchase the right from the conservation group. Is there a price that the conservation group would accept that would allow an industry to pollute? Yes. If the industry offers the conservation group a price which is greater than the value the group places on clean air, it will sell the right to pollute the air. Alternatively, if the government grants the property right to the air to industry is there any price which industry would accept from the conservation group? Again, the answer is yes. If the conservationists pay the industry a price greater than the benefit the industry gets from polluting the air, the industry will sell the property right. It will not matter which group is given the initial property right. If relatively cost-free private bargaining among parties is possible, the air will be put to its most valued use.

Two startling examples of how property-right ownership can lead to cooperation rather than competition come from the National Audubon Society and The Nature Conservancy (Stroup and Baden 1983, 49–50; Baden and Stroup 1981). The Audubon Society owns the Rainey Wildlife Sanctuary, some 26,800 acres of marshlands in Louisiana. Although the primary purpose of the reserve is to serve as home to fifty thousand snow geese and as a wildlife sanctuary for deer, armadillo, muskrat, otter, and mink, the Audubon Society has found it in its interest to also lease the rights to drill for natural gas and to lease grazing land for private herds. The funds received not only go to professional, dedicated, environmentalist, land managers, but also are used for other Audubon programs throughout the nation. The market, you see, made possible the exchange and cooperation between what all too often are polarized and antagonistic interest groups.

The Nature Conservancy serves the interests of its clientele by buying and managing ecologically sensitive areas. For example, at the Mile Hi/Ramsey Canyon Preserve in Arizona, The Nature Conservancy pays for the maintenance of the area by charging visitors for lodging, pet boarding, and tours. The Nature Conservancy has discovered that the more valuable the natural area, the more it fetches in the market and the more it supports the preservation of ecologically sensitive and valuable land elsewhere.

Unfortunately, in many cases this solution is not practical. The costs of bargaining and the number of potential parties will render this solution impossible. An industry would have to bargain with millions of individuals, all of whom may have a unique value of clean air. If the bargaining solution is not a feasible alternative, then we are left with quantity standards on the one hand and carrots (subsidies) and sticks (penalties in the form of taxes) on the other. Both of these solutions require an estimate of marginal benefits and marginal costs.

Estimating the Benefits of an Environmental Resource

Earlier we noted that many environmental groups are becoming enamored of economics and recognize the need to assess benefits and costs. But, as suggested by a recent editorial in *Audubon* magazine, many do so with trepidation and remain skeptical of the results. As the *Audubon* editorialist puts it:

> [A] trap in cost-benefit analysis is that hard, cold numbers cannot define many predictable benefits. The benefit of, say, an air-pollution-control strategy that has a given cost could be an atmosphere in the Grand Canyon which permits the visitor to see across the chasm. The artist, the poet, and the free spirit in all of us can appreciate the awesome experience we derive from that particular place, but how can we attribute a dollar value to the view? Anyone who has fished in a clean stream with a child who caught a trout knows that numbers cannot help compare the experience with the dollar cost of a sewage-treatment plant. (Berle 1988, 6)

The author's concern is heartfelt and real, but in a modern litigious society discussion, compromise, and decision hinge on an adequate measure of the trade-offs involved. To say that it is impossible to place a value on environmental amenities makes certain that those who do so anyway will have their values enter the analysis. Claims that environmental experiences are priceless amount to claims that they are of infinite value. Such claims are without foundation. Economists cite as evidence the fact that people sometimes postpone a fishing trip because something more pressing (i.e., more valuable) appears on their schedule. While it may be impossible to compare the value of catching a trout with the dollar cost of a sewage-treatment plant, it is not impossible to compare catching a trout with its next best alternative. That is, we can and must measure the value of the experience through an assessment of opportunity costs.

What, then, is the value of a forest, a tree, or a stream? Can we put a price on an antelope, elk, or deer? Many claim this question cannot be answered. In fact, many claim it should not be asked. However, these questions are being asked, and wilderness decisions are being made based on the answers. If we are to determine the most desirable level of an environmental resource, the question must be asked—and it must be answered.

To estimate the costs and benefits of clear-cutting a forest or even of removing a single tree, economists turn to the concepts of willingness to pay and willingness to accept. What would an individual be willing to pay to not have the tree cut down, or what sum of money would the individual be willing to accept to compensate him for the removal of the tree?

Think for a moment of the fishing trip of concern to the author of the foregoing quotation. If his business firm offered to pay him one dollar to stay at work for the weekend, we do not expect that he would forgo the trip. If the firm upped the ante and offered a possible promotion or a substantial bonus worth fifty thousand dollars per year, one expects that he might at least reconsider the urgency of the fishing trip. Would he accept the offer? We don't know, of course, but many people would answer yes to this second offer. And when they do they provide the information necessary to establish a range within which the value of the priceless fishing experience must fall. The value is somewhere between one and fifty thousand dollars. The experience is not infinitely valuable—it is not priceless, after all.

Indeed, we can pin the value down much more precisely. If the firm were to start at a dollar and then bid larger and larger amounts eventually the bid would become large enough that most individuals would likely stay on the job. The same bidding game can be conducted with a potential forest clear-cut, or, for that matter, a wilderness encounter with an extra antelope, a herd of elk, or a single grizzly. This methodology has been used to estimate the value (as measured by the willingness to accept or pay) of many environmental commodities from the value of visibility in the Grand Canyon (Schultz *et al.* 1983, 149–169) to the value of water quality at Lake Okoboji in Iowa (d'Arge and Shogren 1989).

Economists also estimate the benefits of wilderness using travel cost methodology. Suppose you drive to a national forest and set up a camp at an established camping area. This camping area can be described by the physical characteristics which surround it. Maybe a lake or a stream is visible from the camping area. To arrive at this camping area, you may have driven past many other campgrounds. These campgrounds did not have the appeal of the one you chose. Maybe those other camping areas had a lake, but no stream. Since you were willing to drive additional miles to be near a stream, the value of that stream is at least as great as the value of the time and the money you spent to get there.

Perhaps farther down the road another camping area exists with not only a lake and a stream, but also a waterfall. Because you were not willing to pay the extra cost (in dollars and time) to travel farther to be near the waterfall, you do not value the waterfall enough to incur the additional expenses to visit it. Willingly or not, you have revealed something about your valuation of a waterfall.

By adding up the values placed on wilderness experiences and other environmental amenities, economists claim to be able to arrive at a collective social valuation. By comparing the costs and benefits, as measured in the market (or through proxy techniques such as those briefly described herein), the economists hope to reveal the trade-offs involved in a way that makes comparison possible. By maximizing the present net value of the discounted stream of benefits and costs, it is said, the public interest is thereby served. But at this point a meaningful economics of the public interest in wilderness and environmental amenities may break down. One problem in public land management, for example, is that it is unclear just who is allowed to interpret the number created by the economist in making a final decision. Furthermore, there is nothing close to an agreed upon metric to develop the monetized value information. In the next section, we find what may be an even more compelling problem for those interested in making informed judgments concerning the allocation of resources to wilderness.[8]

Wilderness Economics and Individualism

AN EPISTEMOLOGICAL CAVEAT[9]

Take a moment and look back at the subtitle of this article, "What's in It for the Individual?" At issue is what is meant by society's interest in wilderness policy and resource management. What, after all, is the public interest anyway?

On the one hand is the view, held by many wilderness advocates and environmentalists, that we are inseparable from a holistic, social, cultural, and natural environment. This view places as much importance on the collective we, as on the atomistic, singular I. This position holds that leaving the fate of the soil and nature to the market would be tantamount to annihilating all organic, holistic, and integrated forms of existence and replacing them with an atomistic and individualistic one.

On the other hand is the view held by many, perhaps most, economists.[10] They (and by implication their analysis earlier described) hold that discussion of a separate public interest apart from the individual members of society borders on mischief if not demagoguery. The focal point of economics is, after all, individual choice.

It is the individual who possesses values, makes choices, and, if given the freedom, takes actions. All group decisions and actions emanate from the collective decisions and actions of individuals. Social goals such as wilderness and wildlife preservation are considered only to the extent that they reflect the collective values or choices of individuals. Since only individuals act, the economists' argument goes, society as an independent organism cannot exist, cannot have values or goals, cannot be the seat of some vague public interest.

The logic of choice contained in economics literature suggests that when confronted by situations of choice, the individual proceeds rationally from a well-defined set of tastes and preferences. Given available alternatives and limited resources, the individual maximizes his or her satisfaction by choosing rationally among them.

The market, it is claimed, efficiently allocates the available resources among the competing individuals, each of whom has a unique set of tastes and preferences. The problem, of course, is that the individual is such a complex set of roles and norms that, at any moment in time (i.e., when confronted with a promotion or a fishing trip), it is difficult to say that a unique ordering of preference exists in the first place.

As individuals we are defined by the set of roles we have chosen, whether freely or through various forms of social, cultural, and economic coercion. But we are social animals as well. Thus we form ourselves into groups—and we freely do so. Groups extend the meaning of the individual and help achieve mutually desired ends.

We have all been confronted with a situation when, a group to which we belong takes an action which conflicts with many of our other roles and norms. The smallest, most localized groups, of course, cannot act in our name without at least the tacit approval we give by remaining in the group. But local groups join with other groups in the community, county, state, and region. Our control, even our votes, become watered down. And because we must belong to many groups, our choices become confused and often contradictory.

Individuals interact through voluntary exchange, and the economics paradigm submits that all such voluntary transactions between rational individuals must be beneficial to all parties. But it is not only individuals who interact. Groups interact. Society provides a set of institutions within which such group interaction is accomplished.

Free markets are one, but only one subset of those institutions. And it is free markets upon which the economists' analysis of the benefits and costs of environmental change are measured and evaluated. The problem is that the concepts of power, collusion, oligopoly, cartels, and general market failure often take on more importance than efficiency and competitive markets in complex group interactions.

The behavior of groups can be effectively analyzed and predicted once it is recognized that the decision makers of the group behave in response to the information available and to the set of incentives which they face. And it is to the analysis and formulation of those incentives that the use of economic analysis is perhaps most appropriate as discussed above. The specific prices established in markets that respond to individuals maximizing behavior is likely to be less important than the fact that trade-offs must be faced and that incentives matter.

Conclusion

If economists can work together with other scientists, an estimation of the marginal benefits and marginal costs of changes in the quality of the environment is possible, and a scheme of taxation and quantity controls can be enacted to achieve a specific desired level of environmental quality. Both local and national wilderness advocates have successfully employed such economic efficiency analysis to prevent wholesale destruction of specific environments and to raise the consciousness of resource administrators.

It is important, however, that those who would use this method remember the underlying epistemological assumption of economics—that only individuals count. The atomistic and individualistic premise that underlies all economic analysis may be at odds with those who claim a holistic, integrated, and synergistic ecological view of the relationships between man and nature. Economics has much to offer, but it cannot deliver the Holy Grail. Nor can economics make our choices; we must do that through political and institutional means.

[Editor's note: Despite the authors' assertion that natural systems tend to reach an equilibrium, several ecologists suggest otherwise (e.g., Botkin, D.B. Discordant harmonies: A new ecology for the twenty-first century. New York: Oxford University Press; 1990.]

Acknowledgment

The authors are Professor and Department Chair of Economics, Weber State University, and Associate Professor of Economics, Weber State University, respectively. Both are Willard L. Eccles Fellows and wish to thank the Eccles Foundation for its generous support of their research. This chapter is a revised version of a paper originally presented under the title "Economic Imperialism and Wilderness Values: A Cautionary View," at the First North American Interdisciplinary Wilderness Conference, Ogden, Utah, February 9–11, 1989.

Notes

1. Peter M. Emerson, "An Overview of the Below-Cost Timber Sales Issue," in LeMaster, D. C.; Flamm, B. R.; Hendee, J. C., eds., *Below-Cost Timber Sales: Proceedings of a Conference*; 1986, February 17–19; Spokane, WA (Washington, D.C.: Wilderness Society, 1987), pp. 1–7. The authors thank Con Schallau for allowing them to incorporate this interpretation of Emerson's comments here; taken from an unpublished outline of a paper prepared by C. H. Schallau, W. R. Maki, and D. Olson, "Some Economic Implications of a Change in Timber Harvesting on the Tongass National Forest," (*n.d.*) The net present value calculation simply attempts to compare the discounted net benefits (benefits less costs over time) of a project against other alternative projects that yield a rate of return equal to the current market interest rate. In its simplest form, this amounts to deciding whether to buy a certificate of deposit from Bank A that yields four percent when Bank B, next door, offers an alternative certificate of deposit that yields six percent.

2. For an in depth review of O'Toole's suggested reforms, which center on the concept of marketization of much public land management, see Richard M. Alston (1989).

3. Con Schallau, a referee for this paper, pointed out that it is important not to leave the impression that there is knowledge when, in fact, much is lacking.

4. Kennedy P. Maize, ed., *Blueprint for the Environment: Advice to the President-Elect from America's Environmental Community.* (Washington, D.C.: Blueprint for the Environment, 1988), pp. 12–13. Included in the list of 18 sponsoring organizations are Defenders of Wildlife, Friends of the Earth, Izaak Walton League, National Audubon Society, National Wildlife Federation, NRDC, Sierra Club, The Wilderness Society, and Trout Unlimited. This is simply one more example, as pointed out at the beginning of the paper, that many environmental groups are coming to believe that using the economist's analysis of efficient resource allocation serves their own self-interest.

5. As one wag has stated the matter, "People ardently opposed to both acid rain and nuclear power may have to decide which bumper sticker to pull off." Andrew Kupfer, "Managing Now for the 1990's," *Fortune* 118(7):45 (April 26, 1988).

6. The classic article on this topic is G. Hardin, "The Tragedy of the Commons," *Science* 162:1243–1248 (1968).

7. The designated owner, of course, may be the government itself, in which case the public at large would charge the specific users of the environmental resources. See, for example, the innovative proposals for "marketization" of environmental resources in O'Toole, *Reforming the Forest Service*, pp. 196–234.

8. The following section suggests an epistemological reason for questioning the value of economists' estimates of opportunity costs. A more direct reason, suggested by David Iverson, a regional economist for the U.S. Forest Service, is that in practice the contingent value and travel cost methodologies do not yield opportunity cost measures that can be compared to other monetary measures. They may, however, be useful in developing a lower-bound estimate for what we value an area to be worth when, for example, we designate it to be wilderness. See J. G. Jones *et al.* (1978: 410–422) for an interesting application of the economics approach that does not claim to be capturing all social values in a single index number.

9. This brief caveat does not do justice to the issues involved. See Alston (1983) for an extended discussion.

10. For an excellent review of the approaches that separate the schools of economic thought on this issue, see Randall (1985).

References

Alston, R. M. The individual vs. the public interest: Political ideology and national forest policy. Boulder, CO: Westview Press; 1983.

———. Reforming the forest service. Ecology Law Quarterly 15(3):503–517; 1989.

Baden, J.; Stroup, R. Saving the wilderness. Reason 13:28–36; 1981.

Berle, P. A. A. Numbers can fool you. Audubon (July):6; 1988.

d'Arge, R. C.; Shogren, J. F. Okoboji experiment: Comparing non-market valuation techniques in an unusually well-defined market for water quality. Ecological Economics 1:251–281; 1989.

Emerson, P. M. An overview of the below-cost timber sales issue. In: Lemaster, D. C.; Flamm, B. R.; Hendee, J. C. eds. Below-cost timber sales: Proceedings of a conference; 1986; Spokane, WA. Washington, DC: The Wilderness Society; 1987:1–7.

Jones, J. G.; Beardsley, W. G.; Countryman, D. W.; Schweitzer, D. L. Estimating economic costs of allocating land to wilderness. Forest Science 24:410–422; 1978.

Kupfer, A. Managing now for the 1990's. Fortune 118(7):45; 1988.

Maize, K. P., editor. Blueprint for the environment: Advice to the President-elect from America's environmental community. Washington, DC: Blueprint for the Environment; 1988.

O'Toole, R. Reforming the forest service. Washington, DC: Island Press; 1988.

Randall, A. Methodology, ideology, and the economics of policy: Why resource economists disagree. American Journal of Agricultural Economics 67:1022–1028; 1985.

Sample, V. A.; Kirby, P. C. National forest planning: A conservationist's guide. Washington, DC. The Wilderness Society, Sierra Club, Natural Resources Defense Council, National Audubon Society, National Wildlife Federation; 1983.

Schultze, W.; Brookshire, D. S.; Walther, E. G.; MacFarland, K. K.; Thayer, M. A.; Whitworth, R. L.; Ben-David, S.; Malm, W.; Molenar, J. The economic benefit of preserving visibility in the national parklands of the southwest. Natural Resources Journal 23:149–173; 1983.

Stroup, R. L.; Baden, J. A. Natural resources: Bureaucratic myths and environmental management. San Francisco, CA: Pacific Policy Institute; 1983.

Preservation Is Not Enough

The Need for Courage in Wilderness Management

THOMAS L. FLEISCHNER

Introduction

THOUSANDS of people over many years have made tremendous efforts to preserve wild lands. And yet, regrettably, less has been accomplished than we like to think. We delude ourselves in assuming that lands will be preserved as wilderness simply because an act of Congress draws new lines on a map. Out there on the real earth, meadows are being trampled, trails are being eroded, campsites are being degraded, water supplies are being polluted, and wildlife is being extirpated. How can this happen in an age of apparent wisdom, sophistication, and sensitivity to wilderness? What can be done to reverse the degradation and truly preserve our precious wild lands?

In attempting to address these questions and provide the beginnings of answers, I use examples from my own experience in the management of the Greater North Cascades Ecosystem (GNCE) of Washington and British Columbia. The same issues, however, are common to many wildlands, such as the Greater Yellowstone Ecosystem, the Colorado Plateau, the High Sierra, and the Southern Appalachians.

Historical Background

European settlers did not see wilderness as something worth saving until their third century of residence on the North American continent. In fact, these settlers actively destroyed wilderness, transforming native forest into pastoral farmland with the passion of the newly converted. Not until settlement had progressed from coast to coast, leaving a tamed landscape in its wake, were the remaining patches of wild nature considered as something of cultural value. The decade of the 1890s, often viewed as a watershed in American history, saw the Western frontier—perennial hope of a better life for so many generations—effectively closed as the westward drive of Euro-American settlement finally spanned the entire continent.

Spearheaded by John Muir, public sentiment for wilderness preservation gained momentum through the next six decades and culminated with the passage by Congress of the Wilderness Act in 1964 (Fox 1981, Allin 1982, Nash 1982). Hampered by a web of political compromise (Allin 1982), the authors of the Act met the daunting challenge of defining wilderness: "A wilderness, in contrast with those areas where man and his works dominate the landscape, is hereby recognized as an area where the earth and its community of life are untrammeled by man, where man himself is a visitor who does not remain" (for full text and explanation of the Act, see The Wilderness Society 1984).

After striving so long toward the goal of preserving wild places, it is no surprise that little thought had been given to what to do with these places after they were legally classified as Wilderness. Two types of wilderness use were defined: resource extraction and recreation. The former included mining, logging and grazing; the latter fishing, hunting, and, increasingly, nonconsumptive activities, such as hiking and nature study. Recreation, as such, was almost nonexistent in wilderness until relatively recently. Since World War II, however, recreation has become a major influence on wildland ecosystems (see Hammitt and Cole 1987), as well as an important economic force in western North America.

Only recently has attention begun to shift away from the strictly political issues of preservation and classification to the more complex question: "What actions should be taken to keep these places wild?" The interdisciplinary art and applied science of wilderness management was born as this question began to be addressed.

The name "wilderness management" itself betrays the ambiguity which obscures our efforts. Webster tells us that *wilderness* derives from the Old English for "place of wild beasts" and implies an uncontrolled state; *management* involves controlling and directing. Within this inherent paradox we have set for ourselves a formidable task.

Nonetheless, we must work toward a solution. The work of a century to allocate public lands as wilderness will have been for naught if we fail to subsequently protect them. Wilderness designation on a map, or in the *Congressional Record*, means nothing if the landscape itself is abused.

Solutions are sought against a backdrop of increasing urgency. Public demand for wilderness recreation outstrips available resources. Officials at the Mt. Baker-Snoqualmie National Forest in Washington State predict that recreational use will more than double in the next forty years, even though we approach the "practical capacity" for human use of the land today (USDA Forest Service 1987).

Records show that more than six thousand people per summer—sometimes over 200 in a single day—visit Cascade Pass, a fragile subalpine area four miles deep into the North Cascades National Park backcountry. The vegetation here has been so damaged by hikers that it became necessary to begin an intensive revegetation effort and construct a stonework "patio" to stabilize the remaining soil.

Recreational use of wilderness has increased dramatically in the past three decades. Historian Roderick Nash (1982) attributed much of the increased demand for wilderness recreation to a trio of revolutions: in equipment, transportation and information. With the advent of nylon, plastics, and other technological developments, exploring wilderness became easier and more comfortable. With this change, hiking became a leisure activity rather than drudgery. The coincident increases in road building and availability of automobiles opened the wilderness to new hordes. Now someone who knows nothing of a wilderness area can pick up a book and learn everything from road access to trail mileages to suggested lunch stops. Climbers can literally learn the location of every handhold and bivouac site. These aids remove much of the mystery and risk from the potential wilderness experience and a major mental block to entering these areas.

All told, technological and intellectual change has made wilderness less intimidating, more inviting, and easier to reach. Americans have responded in droves, and the question must be asked, "What can be done to save our wildlands from being loved to death?"

Current Problems

Across the land, wilderness is under attack. Noncompatible uses, such as mining, invade some areas, while appropriate wilderness activities at inappropriate levels threaten the health of wilderness ecosystems and the sanctity of human wilderness experience. A four-year study of the GNCE's Alpine Lakes Wilderness, for example, indicates that both user numbers and damage to the landscape exceed planned limits (USDA Forest Service 1991). Similar situations abound throughout the GNCE and the entire National Wilderness Preservation System. For this reason Congressman Bruce Vento (1990) recently called for a "revolution in wilderness management."

The integrity of wilderness is in trouble. The whimsy of the political process often subdivides a wilderness ecosystem into a confusing array of incomplete components. In the Greater North Cascades Ecosystem, for example, the morass of administrative designations includes national park, national recreation areas, national forests, designated wilderness areas, and provincial parks, as well as other state, provincial and private lands. All told, the GNCE is managed by a half-dozen agencies from well over a dozen different administrative offices. It comes as no surprise, then, that ancient coniferous forests, alpine highlands, dry rain shadow pine forests, and arid grasslands—all essential components of the ecosystem, and all one "Home to Grizzly Bear"—are not treated as a whole by those who decide their fate. A unified effort to foster an intact wilderness ecosystem is a far cry from the current reality. Meanwhile, habitats of wilderness species are increasingly being fragmented.

Wilderness suffers from the accumulated abuses of many years. As a consequence, land management agencies must too often deal with crisis situations and attempt to find emergency solutions to immediate problems. In a recent critique of Forest Service wilderness management, failure was attributed to inadequate budget and personnel levels, lack of management accountability, and absence of clear management standards (Beum 1990).

Furthermore, some agencies—particularly the Forest Service—tend to manage wilderness as a subset of recreation, even though that is but one of several purposes of wilderness outlined by the Wilderness Act. As a result they overemphasize recreational uses of wilderness (Clark and Buscher 1990). Managing agencies typically fail to realize that wilderness is at least as high a priority as traditional commodity interests such as timber, mining, and grazing. In most wilderness areas, basic ecological surveys have never been conducted nor have careful management strategies been formed. Lack of personnel, time, and funding contribute to this lack of information and commitment.

Wilderness management professionals recently described eight categories of management problems: trail deterioration, campsite deterioration, litter, crowding and visitor conflict, pack stock impact, human waste, impacts on wildlife and fish, and water pollution (Cole, Petersen, and Lucas 1987). Participants at conferences on wilderness management (Frome 1985, Lime 1990) and recreational impact on wildlands (Ittner et al. 1979) raised a similar array of concerns. In a survey of wilderness managers throughout the National Wilderness Preservation System (Washburne and Cole 1983), almost three-quarters of the respondents reported impacts on vegetation in their areas. These managers—representing a great diversity of geographic areas and governmental agencies—described local resource degradation and lack of solitude due to concentrated wilderness use as their most significant management problems. It becomes apparent when examining these problems that wilderness management is primarily a matter of managing people.

As a result of human-caused problems, the spirit of wilderness is also in trouble. When people visit an apparently wild place in hopes of experiencing its healing and restorative powers (Miles 1987), too often they are confronted with the same human problems they were hoping to escape. Social impacts on the one hand—whether disruption by insensitive fellow wilderness users or overregulation by a managing agency—and physical impacts to the landscape on the other, both threaten the potential for direct human experience with wild nature. Both types of impact also threaten nonhuman values of wilderness.

The practice of wilderness management has typically been viewed as a combination of visitor management, such as limiting access, restricting uses, and educating users, and site management, including policies on trails, campsites, fire, and restoration of damaged areas (Hendee, Stankey, and Lucas 1978; 1990). With only a few exceptions wilderness management is, in fact, people management. The effect of human activities on wilderness can be managed directly, through regulation, or indirectly, by influencing behavior. Researchers (Hendee, Stankey, and Lucas 1978; 1990) and at least one agency (USDA Forest Service 1986) agree that indirect methods, such as educating users and limiting access, are preferable to direct regulatory approaches, such as rationing use, because they minimize intrusion into the human experience of wilderness.

The most effective wilderness management tool is public education. Visitors are more likely to respect a place—whether the issue be trampling vegetation, overusing an area, feeding wildlife, or improperly disposing of waste—when they understand these problems. In my own experience, the vast majority of visitors who mistreated the environment did so out of ignorance, not maliciousness. Eradicating that ignorance is finally being seen as a top priority for wilderness managers (Hansen 1990, Passineau 1990). In its official policy on wilderness management the Forest Service acknowledges that educational approaches are "the primary tools" (USDA Forest Service 1986). Former Chief of the U.S. Forest Service, R. Max Peterson (1985), states that wilderness management is "80–90 percent education and information."

In a national survey of wilderness managers the majority of respondents thought that personal contact with visitors was the most effective management technique (Washburne and Cole 1983). One of the most important responsibilities of wilderness rangers is to make positive educational contacts with visitors. Such contacts range from short, informal chats to publicized interpretative talks. In many cases, educational effort is concentrated at trailheads or information centers. When permits were abolished in the North Cascades' Glacier Peak Wilderness Area a few years ago, the worst effect noticed by the staff was the lost opportunity to talk with visitors about their potential impact.

Most educational efforts thus far have occurred in the wilderness areas themselves, or at their portals. But if wilderness is to remain wild, the public must be educated before it ever arrives in the backcountry (Hansen 1990, Passineau 1990). The GNCE, in an encouraging trend, has developed an extensive wilderness education outreach program for the region's schoolchildren, conducted by the nonprofit North Cascades Institute, with the support of the U.S. Forest Service.

One of the most basic management questions concerns access. Should everyone who wants to visit a wilderness area be allowed to do so? If not, how do we limit numbers? Should visitors camp wherever they choose, or only in designated areas? The use of permits has been a controversial issue. Proponents claim that permits are a necessary management tool (Hendee and Lucas 1973), while others argue vehemently against them, believing they engender a "police-state wilderness" (Behan 1974).

In practice, adjacent areas have vastly different policies. North Cascades National Park has one of the most restrictive and regulated backcountry visitation policies in the nation. To obtain a permit, visitors must provide, in advance, specific details of their hiking activity. Campers must use designated campsites, and they cannot change plans in midtrip unless they encounter a ranger. And they often *do* encounter rangers in the backcountry, much to the chagrin of some visitors. To make this system work, a team of a dozen or so rangers patrols the backcountry using two-way radios, as well as using an information/permit station outside the backcountry. Detractors call this degree of regulation overzealous, saying it ruins their experience of wilderness. Detractors and proponents agree, however, that physical damage to the landscape has decreased since the system was instituted a decade ago. By contrast, adjacent national forest wilderness areas in the North Cascades have an "open-door" policy, requiring no permit and no contact with agency personnel. This approach appeals to visitors who cringe at institutionalization of wilderness. But, unarguably, many of these unregulated lands have seriously deteriorated. Over twenty years of monitoring the impact of backcountry recreation on the subalpine vegetation of the North Cascades indicates clearly that unrestricted recreation leads to ecological deterioration of popular sites (Thornburgh 1986; 1990).

Much emphasis has been given to management of wilderness campsites, as these are where physical and social impacts are concentrated. Controversy has existed over the relative virtues of concentrating and dispersing camping use, as well as the effectiveness of closing impacted areas altogether. Concentrating campsite use appears to be superior to visitor dispersal for minimizing physical impact throughout a wildland (Cole 1981). Because most impact occurs within the first few years of use, campsite closures are often ineffective (Cole 1981, Cole and Marion 1986). A variety of techniques for monitoring impact of wilderness campsites has been developed (Cole 1983; 1989). Cole (1990b) has put forth a useful set of campsite management principles. Human waste in high-use areas creates another web of problems. To cope with this challenge, latrines are often placed in high-use areas and campsites. These toilets range from primitive pits to experimental composting models in use in North Cascades National Park (Weisberg 1988).

Sometimes the effort to preserve wildness leads to paradoxical practices. At Cascade Pass, in an ongoing management activity, we shoveled hiking routes across early season snowfields, attempting to align the snow trail as closely as possible to the true trail below. Hundreds of trampling feet could then melt out on the trail rather than on fragile meadows. But the image of a uniformed employee shoveling snow below glaciers left some visitors doubting the wildness of the place.

In a similar move to preserve fragile vegetation, we placed agricultural salt blocks for deer in a subalpine basin. Paradoxically, this was intended not to tame the deer but to keep them wild. Salt is a limiting resource in the subalpine environment of the North Cascades, and all nonhuman mammals are drawn to it. The most abundant sources of salt are provided by humans through urine and sweat. Deer and other mammals stay around campsite areas specifically to obtain salt and, in the process, become corrupted by food handouts from uninformed hikers. Our salt blocks were an attempt to wean them from this self-destructive addiction.

Increasingly, managers are focusing on restoration and "repair" of damaged areas (Pollock 1988, and Ouderkirk, this volume). An innovative and successful revegetation project at North Cascades National Park (Lester and Calder 1979) has inspired similar work elsewhere. Previously denuded subalpine meadows have been revegetated with native species grown from seed in a greenhouse and planted in their area of origin. This time- and labor-intensive procedure often yields dramatic results which are conspicuous to returning backcountry visitors. In a positive (but too rare) example of interagency cooperation, the National Park Service now helps train Forest Service personnel in these revegetation techniques.

Planning for the future of wilderness has generally been haphazard or nonexistent. Managers' time and energy are often stretched too thinly with present matters to engage in the time-consuming task of planning. Recently, however, Forest Service researchers developed a planning system, which can be used in a variety of ecosystems. Called Limits of Acceptable Change, or LAC (Grumbine 1985, Stankey et al. 1985, Wuerthner 1990), the system makes two major advances. First, it provides a process for determining the desirable conditions for an area, rather than the merely tolerable. Second, it provides managers with an existing planning procedure, hopefully speeding up the planning process.

LAC is currently the primary focus for agency managers involved in wilderness planning. Prominent conservation analyst Michael Frome (personal communication) feels that LAC falls well short of its goals by allowing managers to avoid decisions by using a cumbersome bureaucratic procedure. While success or failure remains undetermined, LAC clearly signals that managers are now recognizing the need for long-range planning and for establishing limits on cumulative impacts on wildlands.

No system of long-range planning, however, will succeed without gathering more accurate information on the biological, physical, and social environments to serve as a baseline to monitor change (Hendee, Stankey, and Lucas 1990, Cole 1990a, Stankey 1982). Simply put, it is hard to get somewhere if you don't know your starting point. If planning is to ever bear fruit, managing agencies must place baseline resource and use inventories as higher priorities.

Several authors have proposed sets of principles for managing wilderness. Hendee, Stankey, and Lucas (1978) set forth the first set of principles over a decade ago, and their work dramatically influenced managing agencies. The Wilderness Society (1984) and Reed, Haas, and Beum (1990) augmented the original, solid work of Hendee and his colleagues with additional principles for wilderness management, while Hendee, Stankey, and Lucas (1990) recently made their own revisions. Taken as a whole, these guidelines offer insightful direction to managers. The trend over the years toward recognizing biocentric values and the importance of maintaining ecosystem integrity is encouraging. Nevertheless, more sweeping changes in wilderness management are called for.

The Changes Called For

In spite of the genuine concern and painstaking work of dedicated wilderness managers, the approaches tried thus far are inadequate and doomed to failure unless major changes are made. Cole (1990a) pointed out that changes must occur both inside and outside the managing agencies. He urges agencies to expand research programs, encourage planning and monitoring, upgrade the training of personnel, and increase the accountability of managers. He encourages those outside the agencies—academicians, conservation groups, and the general public—to demand professional management and to become more involved in wilderness management issues. A recent critique of Forest Service wilderness management saw the need for increased management accountability, clarified resource standards, and increased funding and personnel (Beum 1990).

Several other fundamental changes are called for, if indeed we are to maintain wild nature even as we mingle with it.

BECOMING VISIONARY

Wilderness management today suffers from a lack of vision. Long-term planning for wilderness is in its earliest stages. We, as a society, have not yet begun to take the even longer view necessary. We should be planning for ecosystem health not just years, but centuries and millennia from now. Wilderness managers must ask: What should this place look like in one hundred years? What type of experience can a hiker have on this ridge two hundred years from now?

To keep our wilderness vision broad we must spend as much time as possible *out there*. Wilderness managers often become too bogged down in paperwork and embroiled in contemporary brushfires to maintain an expansive view of the function and power of wilderness and of the potential of their work. Often those with the most power to affect an area know it least intimately. Decisions made in offices closed away from the outdoor reality seldom speak loudly or clearly enough for the needs of the place. None of the headquarters and planning offices of the federal agencies that manage the Greater North Cascades Ecosystem lie within their respective boundaries. It should be an agency requirement that all managers and bureaucrats spend at least two weeks a year in any wilderness whose future they help shape. Responsibility to the place itself—rather than to political powers—is felt more clearly while sleeping on the ground and listening to its spirit than while sitting in a windowless meeting room, behind a computer, or at the end of a telephone cord. The General Accounting Office recently reported that 130 Forest Service ranger districts with wilderness responsibilities did not have a single employee who spent even ten percent of his/her time on wilderness management (Vento 1990).

Part of the cause for lack of vision lies in the inherent nature of bureaucracies. Most wilderness managers work for the federal government—the biggest bureaucracy of all. Agency workers often feel defeated as they see their hard work cast aside for political reasons. Unsympathetic attitudes of presidential administrations quickly filter down to field offices in the hinterlands. Progress moves interminably slowly at times as the wheels of bureaucracy turn. Promotions are generally not based on who has been the most committed to the long-term health of a wilderness ecosystem. For the sake of their own survival, managers avoid taking strong actions. In such a social milieu, committed workers too often become demoralized and mediocrity flourishes.

Wilderness management provides a classic example of the superiority of "an ounce of prevention over a pound of cure." Many management problems are the direct result of having wilderness areas that are too small to support self-sustaining ecological processes. The lack of sufficiently large areas, in turn, follows from a lack of vision in the political process of wilderness designation. Wilderness managers need to become more involved in the controversial politics of designation to advocate "big wilderness" (Foreman 1991). Large areas are more capable of managing themselves, saving both habitat and taxpayer dollars.

REDUCE ACCESS

One of the most tangible means of managing wilderness wisely is one of the simplest: close roads. By reducing road access into wilderness we vastly increase its size and integrity. The passage into wilderness is as important a part of the backcountry experience as the arrival at a destination. Managers do a disservice to hikers by allowing overly easy access to the high country; they further the illusion that one can bond with wilderness by racing in and out. In the North Cascades, climbers routinely rush to the high country to bag a peak, then rush back out, cursing any forest they must travel through. Managers err seriously by promoting this attitude and behavior. Rather, our goal should be, as Joseph Sax (1980) put it, "to engage the contemplative faculties" of visitors. We can dramatically reduce physical impact on the environment, and deepen the quality of a visitor's experience by reverting access roads to hiking trails.

This wilderness, the last we have, cannot be all things to all people. There is absolutely no responsibility to provide easy auto access to dramatic viewpoints. This only promotes a postcard view of nature—nature *looked at* rather than *lived in*. Moreover, with most of the United States already containing extensive roads, the opportunities for vehicular exploration far outstrip the availability of genuine wilderness experience. We have the knowledge and technology to restore roads to trail conditions. All that we lack is the wisdom.

ACHIEVING TRUE BIOCENTRISM

Wilderness managers should be leaders in cultural transformation. For long-term sustainable human culture, we must move from anthropocentrism (human-centeredness) to biocentrism (life-centeredness). In seeking biocentrism we should follow the direction of Devall and Sessions (1985), who note that "all things in the biosphere have an equal right to live and blossom and to reach their own individual forms of unfolding," rather than the distorted view of earlier wilderness managers (Hendee and Stankey 1973), who implausibly described biocentrism in terms of human benefit. When we spend time living in wild places we see clearly that wilderness is more than a human playground; it is home for myriad forms of life.

Restoration of damaged areas is gratifying work, but several cautions are in order. First, it is far easier to *avoid* problems than to repair them. Second, in many cases, vegetation cannot be replaced with the original species because oftentimes those original species cannot be propagated. At Cascade Pass, original site of restoration efforts in the North Cascades, showy sedge (*Carex spectabilis*) and partridge-foot (*Leutkea pectinata*) now predominate in many areas originally covered by heathers (*Phyllodoce empetriformis* and *Cassiope mertensiana*). These new meadows stand as a tribute to years of dedication and hard work, and are vastly superior to bare ground, but they are not a restoration of the original plant community. Lastly, the public should not be led to believe that managers can fix anything that visitors destroy. If this attitude were to become pervasive, restoration efforts would backfire: rather than undoing previous damage, they would encourage more damage. Replanting damaged meadows should never become a routine part of an agency's work.

ENCOURAGING INDIVIDUAL RESPONSIBILITY

As with restoration, so with all of wilderness management: individual responsibility must be encouraged. Wilderness managers should be neither Big Brother nor janitor. Ultimately, the goal of the wilderness manager should be to work oneself out of a job by encouraging the public in every way to share in the responsibility of taking care of these precious lands. The agencies cannot, and should not, tend to hikers like flocks of errant sheep.

GREATER EDUCATIONAL EFFORT

Of the many steps necessary to transform our relationship with wilderness, none is more fundamental than education. Agencies must transcend their political boundaries and interact with users *before* they enter wilderness. Education and awareness of nature, including wilderness values, should be an integral part of every school's curriculum (see Charles, this volume). Management agencies should aggressively promote this idea and offer their services to schools. Interpretive programs in park and forest areas should always be top budget priorities.

Obviously, these changes will not occur overnight. One thing, however, is certain: if we don't try, we won't succeed.

The following steps are necessary for change:

1. Establish a vision of the ideal.
2. Take decisive action toward that goal, even if such action is controversial and unpopular with some of the public.
3. Play a strong and conspicuous role NOW—closing roads, limiting numbers of visitors—with the intent of minimizing our role in the future.
4. Approach decisions biocentrically, never forgetting that the integrity of nonhuman lives is of equal concern as our own lifestyle.

Wilderness represents one extreme of the world's land-use spectrum. Over ninety-five percent of the contiguous United States has been altered from its original wilderness state. Only two percent is legally protected from exploitative uses. To even consider allowing further degradation of this tiny remnant of wild land is a travesty, not to mention a violation of the letter and spirit of the Wilderness Act.

Our ultimate goal is to rediscover old ways of living lightly upon the landscape, to again find the place where wilderness and civilization are compatible. We must intermingle more freely with wild nature; we must learn from it without destroying it. For this possibility to become reality, we must take strong action now. Do we have the courage to take the first step?

Acknowledgments

Some of these ideas appeared in different form in *Forever Wild: Conserving the Greater North Cascades Ecosystem*, M. Friedman, ed. 1988. Mountain Hemlock Press, Bellingham, Washington.

My thoughts on wilderness and its management have been deepened by dialogue with many people over many years. I would especially like to thank Edie Dillon, Ed Grumbine, and Saul Weisberg for sharing their insights. My students at the Sierra Institute of the University of

California-Santa Cruz and Prescott College helped sharpen my thinking during many backcountry discussions. And most of all, I thank the North Cascades and the desert canyons of the Southwest, my two most profound teachers.

References

Allin, C. W. The politics of wilderness preservation. Westport, CT: Greenwood Press; 1982.

Behan, R. W. Police state wilderness. Journal of Forestry 72:98–99; 1974.

Beum, F. R. A time for commitment: Case-study reviews of national forest wilderness management. Washington, DC: The Wilderness Society; 1990.

Clark, R. N.; Buscher, R. F. Managing lands adjacent to wilderness. In: Lime, D. W., ed. Managing America's enduring wilderness resource. St. Paul: Minnesota Extension Service, University of Minnesota; 1990:440–445.

Cole, D. N. Managing ecological impacts at wilderness campsites: An evaluation of techniques. Journal of Forestry 79:86–89; 1981.

―――. Monitoring the condition of wilderness campsites. Ogden, UT: U.S. Dept. of Agriculture Forest Service Research Paper INT-302; 1983.

―――. Wilderness campsite monitoring methods: A sourcebook. Ogden, UT: U.S. Dept. of Agriculture Forest Service General Technical Report INT-259; 1989.

―――. Wilderness management: Has it come of age? Journal of Soil and Water Conservation 45:360–364; 1990a.

―――. Some principles to guide wilderness campsite management. In: Lime, D. W., ed. Managing America's enduring wilderness resource. St. Paul: Minnesota Extension Service, University of Minnesota; 1990b: 181–187.

Cole, D. N.; Marion, J. L. Wilderness campsite impacts: Changes over time. In: Lucas, R. C., comp. Proceedings—national wilderness research conference: Current research. Ogden, UT: U.S. Dept. of Agriculture Forest Service General Technical Report INT-212; 1986:144–151.

Cole, D. N.; Petersen, M. E.; Lucas, R. C. Managing wilderness recreation use: Common problems and potential solutions. Logan, UT: U.S. Dept. of Agriculture Forest Service General Technical Report INT-230; 1987.

Devall, B.; Sessions, G. Deep ecology: Living as if nature mattered. Salt Lake City, UT: Peregrine Smith Books; 1985.

Foreman, D. Dreaming big wilderness. In: Confessions of an eco-warrior. New York: Harmony Books; 1991:177–192.

Fox, S. The American conservation movement: John Muir and his legacy. Madison: University of Wisconsin Press; 1981.

Friedman, M., editor. Forever wild: Conserving the Greater North Cascades ecosystem. Bellingham, WA: Mountain Hemlock Press; 1988.

Frome, M., editor. Issues in wilderness management. Boulder, CO: Westview Press; 1985.

Grumbine, R. E. Can wilderness be saved from Vibram soles? High Country News. 1985 May 27:14–15.

Hammitt, W. E.; Cole, D. N. Wildland recreation: Ecology and management. New York: John Wiley and Sons; 1987.

Hansen, G. F. Education, the key to preservation. In: Lime, D. W., ed. Managing America's enduring wilderness resource. St. Paul: Minnesota Extension Service, University of Minnesota; 1990:123–130.

Hendee, J. C.; Lucas, R. C. Mandatory wilderness permits: A necessary management tool. Bioscience 23:535–538; 1973.

Hendee, J. C.; Stankey, G. H. Biocentricity in wilderness management. BioScience 23:535–538; 1973.

Hendee, J. C.; Stankey, G. H.; Lucas, R. C. Wilderness management. Miscellaneous Publication No. 1365. 1978. U.S. Dept. of Agriculture Forest Service, Washington, DC.

———. Wilderness management. Revised ed. International Wilderness Leadership Foundation. Golden, CO: Fulcrum Publishing; 1990.

Ittner, R.; Potter, D. R.; Agee, J. K.; Anschell, S.; editors. Recreational impact on wildlands: Conference proceedings. 1979. Available from: U.S. Dept. of Agriculture Forest Service, Pacific Northwest Region, Portland, OR.

Lester, W.; Calder, S. Revegetating the forest zone of North Cascades National Park. In: Ittner, R.; Potter, D. R.; Agee, J. K.; Anschell, S., eds. Recreational impact on wildlands: Conference proceedings; 1979:271–275. Available from: U.S. Dept. of Agriculture Forest Service, Pacific Northwest Region, Portland, OR.

Lime, D. W., editor. Managing America's enduring wilderness resource. St. Paul, MN: Minnesota Extension Service, University of Minnesota; 1990.

Miles, J. Wilderness as healing place. Journal of Experiential Education 10(3):4–10; 1987.

Nash, R. Wilderness and the American mind. 3d edition. New Haven, CT: Yale University Press; 1982.

Passineau, J. F. Teaching a wilderness ethic: Reaching beyond the forest. In:

Lime, D. W., ed. Managing America's enduring wilderness resource. St. Paul: Minnesota Extension Service, University of Minnesota; 1990:133–141.

Peterson, R. M. National Forest dimensions and dilemmas. In: Frome, M., ed., Issues in wilderness management. Boulder, CO: Westview Press; 1985:36–52.

Pollock, S. A time to mend. Sierra 73(5):50–55; 1988.

Reed, P.; Haas, G.; Beum, F. Management principles for a 1990's wilderness revolution. In: Lime, D. W. ed. Managing America's enduring wilderness resource. St. Paul: Minnesota Extension Service, University of Minnesota; 1990:250–258.

Sax, J. L. Mountains without handrails: Reflections on the national parks. Ann Arbor: University of Michigan Press; 1980.

Stankey, G. H. The role of management in wilderness and natural-area preservation. Environmental Conservation 9:149–155; 1982.

Stankey, G. H.; Cole, D. N.; Lucas, R. C.; Petersen, M. E.; Frissell, S. S. The limits of acceptable change (LAC) system for wilderness planning. Ogden, UT: U.S. Dept. of Agriculture Forest Service General Technical Report INT-176; 1985.

Thornburgh, D. A. Responses of vegetation to different wilderness management systems. In: Lucas, R. C. comp. Proceedings—national wilderness research conference: Current research. Ogden, UT: U.S. Dept. of Agriculture Forest Service General Technical Report INT-212; 1986:108–113.

———. Success of a "gentle persuasion" wilderness management system. In: Lime, D. W., ed. Managing America's enduring wilderness resource. St. Paul: Minnesota Extension Service, University of Minnesota; 1990: 121–122.

United States Department of Agriculture Forest Service. Forest Service Manual, Chapter 2320, Wilderness management. 1986. Available from: U.S. Dept. of Agriculture Forest Service, Washington, DC.

———. Proposed land and resource management plan, Mt. Baker-Snoqualmie National Forest. 1987. Available from: U.S. Dept. of Agriculture Forest Service, Mt. Baker-Snoqualmie National Forest, Seattle, WA.

———. Alpine Lakes Wilderness recreation visitor use monitoring report. 1991. Available from: U.S. Dept. of Agriculture Forest Service, Wenatchee National Forest and Mt. Baker-Snoqualmie National Forest, Seattle, WA.

Vento, B. F. A wilderness revolution for the 1990's. In: Lime, D. W., ed. Managing America's enduring wilderness resource. St. Paul: Minnesota Extension Service, University of Minnesota; 1990:9–17.

Washburne, R. F.; Cole, D. N. Problems and practices in wilderness management: A survey of managers. Ogden, UT: U.S. Dept. of Agriculture Forest Service Research Paper INT-304; 1983.

Weisberg, S. Composting options for wilderness management of human waste, North Cascades National Park Service Complex. 1988. Available from: U.S. Dept. of the Interior National Park Service, North Cascades National Park Service Complex, Sedro Woolley, WA.

The Wilderness Society. The Wilderness Act handbook. Washington, DC: The Wilderness Society; 1984.

Wuerthner, G. Managing the "Bob." Wilderness, Spring 1990: 45–51; 1990.

Future

Paths

Wilderness

Promises, Poetry, and Pragmatism

JAY D. HAIR

ACROSS the country, the debate about what to designate as wilderness is one of the most dramatic arguments on the environmental front. The incalculable value of our natural legacy, the grandeur of wilderness, the economic impact of protection are all elements in that debate. That is why I have titled this paper "Wilderness: Promises, Poetry and Pragmatism." Let me look at each of those aspects one at a time. First the promises.

Promises

By the time this nation turned its ingenuity to the *protection* of wilderness, it had nearly completed its *conquest* of wilderness. American pioneers, like people since Biblical times, fought the "wilderness" in a relentless drive for civilization. In a way they succeeded: when the 1890 United States census was taken, the frontier land was officially declared "gone."

Only in the last few decades—in the decades of Bob Marshall, Olaus Murie, Aldo Leopold and others—have we at last realized that wilderness *is* a vital part of civilization. We finally comprehended that, as we had conquered the wilderness, we had really destroyed a unique component of our natural heritage. By the 1960s, we fully realized our loss and began to accept our responsibility to protect the few wild places still untouched across our nation. So began the long and successful drive for passage of The Wilderness Act.

The Act was rewritten sixty-six times before it was finally passed by Congress and signed by President Lyndon Johnson on September 3, 1964. Idaho's late Senator Frank Church was the Congressional floor manager for this landmark piece of legislation.

In passing the Wilderness Act, Congress sought to assure citizens that "an increasing population, accompanied by expanding settlement and growing mechanization, does not occupy and modify all areas within the United States." The legislation attempted to "secure for the American people of present and future generations the benefits of an enduring resource." And it recognized the value of areas "where the Earth and community of life are untrammeled by man—where the imprint of man's work is substantially unnoticeable."

The goals were lofty. The promises were broad. But have they been kept?

Today, the National Wilderness Preservation System contains over 91 million acres of world-class resources. It is the largest system of wild lands protection in the world, and the envy of other nations.

Nonetheless, wilderness covers less than four percent of the United States' total land mass. It is far less than we need. And it is far less than we promised to the American people when the Wilderness Act was signed. But even as we acknowledge that, we must ask: Why wilderness? Why should we maintain areas where humans and their works *do not* dominate the landscape? In the title, I promised you poetry, and in poetry we find some answers.

Poetry

But I'm not a poet. I'm a scientist and conservationist. So let me use the words of another who has captured America's deep need and love for wild places and things. Robert W. Service, the English-born Canadian poet, says the following in one of his best known poems, "The Call of the Wild" (Service 1954):

Have you gazed on naked grandeur where
    there's nothing else to gaze on,
Set pieces and drop-curtain scenes galore,
Big mountains heaved to heaven, which the
    blinding sunsets blazon,
Black canyons where the rapids rip and roar?
Have you swept the visioned valley with
    the green stream streaking through it,
Searched the Vastness for a something you have lost?
Have you strung your soul to silence? Then
    for God's sake go and do it;
Hear the challenge, learn the lesson, pay the cost.

Have you wandered in the wilderness, the sagebrush
desolation, . . .
Have you camped upon the foothills, have you galloped
o'er the ranges,
Have you roamed the arid sun-lands through and
through?
Have you chummed up with the mesa? Do you know its
moods and changes?
Then listen to the Wild—it's calling you.

And the last verse has a message for all of us:

They have cradled you in custom, they have
primed you with their preaching,
They have soaked you in convention through and through;
They have put you in the showcase; you're a credit to
their teaching—
But can't you hear the Wild?—it's calling you.
Let us probe the silent places, let us see what luck
betide us;
Let us journey to a lonely land I know.
There's a whisper on the night-wind, there's
a star agleam to guide us,
And the Wild is calling, calling . . . let us go.

In fact, statistics indicate that more and more of us are answering
the call of the wild. Why? Because in the wild, we find ourselves. In the

wilderness, we find our links to yesterday and to eternity. Nonetheless, we still haven't done enough to protect that heritage. When it comes to the issue of wilderness, we, like the renowned American poet Robert Frost, "have promises to keep and miles to go before [we] sleep."

Pragmatism

So far I've given you the promises and the poetry of wilderness. Now it's time for pragmatism. It's time to discuss the problems we face in a world growing more complex with each passing day. And it's time to expand the scope of this presentation.

Every environmental issue needs to be put into a larger perspective. While the focus of a given issue may be local, we must always remember that it is just one part of a set of global concerns.

So let me put the issue of wilderness into its global context: We live in a world where an unprecedented number of people are well fed, well clothed and well housed. Yet, we also live in a world where as many as one hundred thousand people starve to death each day.

We live in a world of opulence, a world where a Japanese firm had spent 40 million dollars for one Van Gogh painting. Yet, we also live in a world where more than 800 million people live in conditions the World Bank describes as "absolute poverty—life degraded by disease, illiteracy, malnutrition and squalor."

We live in a world in which we consume well over a third of total terrestrial photosynthetic productivity (Raven 1986). And we live in a world in which, for the first time in the history of civilization, every human being is in contact with potentially dangerous chemicals from the moment of conception to the time of death.

The complexities and contradictions of contemporary society are evident in other ways as well. For example, in recent years, society has made stunning technological advances in medicine, space exploration, global communication systems, and agricultural productivity. Our learning curve is so advanced that, at any given moment, we can measure the distance between Earth and the moon—which is almost a quarter of a million miles—and be off by less than half an inch. That is an amazing accomplishment. Yet, what we *don't* know is even more amazing.

For instance, we don't know how many species of life share this planet with us humans. According to E. O. Wilson (1985) of Harvard University, "We do not know, even to the nearest order of magnitude."

We know that about 1.7 million species have been formally named since Linnaeus inaugurated the binomial system of scientific nomenclature in 1753. In the 1960s and 1970s, a few scientists estimated the world's total number of species as high as 10 million. Then, in 1982, after an intensive sampling of tropical rain forests, others raised the estimate by threefold.

So how many species live on Earth? The answer is still a mystery, and it has a direct relationship to our need for protected wilderness ecosystems.

Because, of even greater concern, is scientific evidence that we are witnessing the global destruction of world-class wilderness ecosystems—particularly tropical forests. If unchecked, that process will culminate in the summary elimination of millions of species. Norman Myers (1986) noted, "Of all the environmental assaults we are mounting against the Earth, mass extinction will be the most profound."

Isn't it ironic that just when we are learning so much about the origins of life, we are also allowing so much of life's biological diversity to disappear? Isn't it tragic that just when we are learning how to improve the quality of life through spectacular advances in bioengineering and associated technologies, we are also allowing entire stocks of genetic materials to be eliminated?

Those are elements of the global picture. But how do they relate to the protection of wilderness areas in North America? To some, it may seem a tenuous connection. To others, it may appear irrelevant. It may seem like just another academic question. I hope to convince the skeptics otherwise. Because our need for wilderness is more than just aesthetic, more than just spiritual, and more than just poetic. Our need for wilderness, perhaps most of all, is scientific and economic.

Let us first examine the scientific values. As the basic unit of evolutionary biology, the species is also the basic unit of ecology. An ecosystem, comprised of species in association with their environments, is best understood when we can divide it into its component parts.

Then we can understand the relationships within and between species and their habitats. If we do not have intact natural ecosystems—such as those found in large, undisturbed wilderness areas—, then we severely limit our global—and our local—opportunities for studying the determinants of species diversity, population regulation, energy cycles, nutrient flows, social systems, and community structure.

All are critical to understanding how natural ecosystems function—whether it is the relationship of elk to their habitat in Idaho or the relationship of humankind to the biosphere.

Dr. Maurice Hornocker, associated with the University of Idaho and the National Wildlife Federation, showed us that in seminal research which he and his colleagues conducted on mountain lions in central Idaho. His long-term research project demonstrated the relationships between mountain lions' intrinsic behavior and the wilderness areas they inhabit. Most important, his research highlighted the fact that knowledge about the effects of species on their habitat is essential if we are to make sound management decisions for the future of any ecosystem. Too little research of this type is being conducted today, and we need to understand why.

After all, most in society agree that scientific inquiry is essential if we are to understand the world around us. And most in society agree that such knowledge is highly valuable in formulating solutions to resource management problems. Therefore, why has so little long-term research about appropriate wilderness-related topics been undertaken? Partially because many in our society can't see the importance of wilderness eco-systems until they are shown their economic values.

So let me do just that—first on a global scale and then with an example from within the United States.

Worldwide, every time a prescription drug is bought, there is a 50-percent chance that the purchase owes its origin to materials from wild organisms. In the United States, the annual commercial value of these medicines is approximately 14 billion dollars. Around the world, the commercial value tops 40 billion dollars a year.

In other words, the pharmaceutical industry has an enormous stake in the health of worldwide wilderness ecosystems. If the current rate of global habitat destruction and species loss continues, the pharmaceutical industry—and humankind—will be denied opportunities to discover new drugs to end the suffering and death of millions.

A Congressional advisory group found that species are disappearing at a rate perhaps not seen since the loss of the dinosaurs 65 million years ago. At this rate, an average of 100 species may become extinct *each day* by the turn of the century. Most extinctions will take place in tropical wilderness areas—like those in Madagascar.

The Madagascar forests, for instance, are the native habitat of the rosy periwinkle. That species contains alkaloids that have yielded two potent medicines against a variety of blood-related cancers. To date, more than ninety-three percent of Madagascar's forests—including habitat for rosy periwinkles—have been destroyed. More than half the native plant and animal species are presumed lost.

The National Cancer Institute has reported that in the Amazon Basin alone there are undoubtedly several other species of plants that could yield "superstar" drugs against cancer. But we may never even know their names because, sadly, as the world loses wild things and wild places, we also lose the myriad benefits they have held secret from humankind.

Now, let's turn to a state in the American West—Idaho. People come to Idaho for a lot of good reasons. For the breathtaking scenery, the forests, the sparkling trout streams, and the wild, untouched lands. They come to Idaho for what its residents already know about. They come for Idaho's spectacular outdoors—they come for the wilderness.

The National Wildlife Federation and our affiliate, the Idaho Wildlife Federation, recently asked 11 thousand nonresident hunters about Idaho's public lands. An overwhelming eighty-seven percent supported the designation of *more* wilderness in the state. Less than five percent of the respondents opposed more wilderness designation. At the same time, more than sixty-eight percent of the hunters said they were satisfied with their hunting experience last year.

What does the survey really illustrate? That Idaho's natural amenities, while important for their aesthetic values, are also important for their tourist and economic potential. The hunters we surveyed will be back next year—and the year after—to pursue recreation in Idaho. So will thousands of others. And they will all bring their checkbooks.

The tourism industry in Idaho has become the state's leading employer, and now rivals traditional industries such as agriculture and mining in overall economic impact. For example, during the 1984–85 outfitting season, nearly seventy thousand people hunted, fished, skied, mountaineered, and otherwise took advantage of Idaho's outdoor resources.

They spent more than 19 million dollars in outfitting and guide activities. Of that amount, nearly 15.5 million dollars stayed in the state. Additionally, the outfitting and guide activities stimulated 24 million dollars in adjunct services for a grand total of more than 38 million dollars poured into the state's economy. The activities created more than 700 full-time jobs.

The 1984–85 season was a record-breaker for the recreation industry in Idaho. And the trend should continue, in Idaho and elsewhere. In the Pacific Northwest, it is estimated that even if all recommended wilderness areas were designated as such, demand for recreation by the year 2030 would still exceed the region's capacity by fifty percent. The economic potential in wilderness is enormous.

The pragmatic—or economic—value of wilderness has not settled the debate about wilderness. The search for a balancing of priorities continues. It is little wonder, because the questions surrounding wilderness are thorny.

In our multiple-options society, how do we provide enough timber to meet our nation's needs while increasing the size of our wilderness system? How do we decide between resources needed for "national security"—resources like minerals, oil and gas—and the resource of land, which warrants protection for its wilderness values?

How do we meet the ever increasing demand for dispersed wilderness recreation—for hunting, fishing, backpacking, rafting—and still maintain the solitude that is central to the "wilderness experience?"

Developing Responsible Options

Do I know the answers? No. I'm not even sure I know all of the questions. However, let me offer some thoughts about how to develop responsible options.

First, we need long-term and properly funded research to provide the kinds of information required to understand complex ecosystems and to resolve complicated public policy and resource management issues.

Let me give you a "real-life" example of how the lack of such long-term research has produced a massive, national environmental conflict. In 1980, when the Alaska National Interest Lands Conservation Act became law, its Section 1002 set aside 1.5 million acres of the Coastal Plain of the Arctic National Wildlife Refuge for further study of its natural resources and potential for oil and gas development.

Although the Coastal Plain—or the 1002 area, as it is commonly called—is a relatively small part of this 19-million-acre wildlife refuge, it is considered the most biologically productive area. It includes the primary calving ground for the internationally invaluable Porcupine caribou herd. This area, as part of an undisturbed arctic ecosystem, is of world-class stature. In fact, the land adjacent to it to the east and south has already been designated as "wilderness."

Now, with virtually no comprehensive research data in hand, the U.S. Department of the Interior is proposing that the entire area be made available for leasing and full, oil-field development. Congress faces two diametrically opposed pieces of legislation: one for total wilderness designation and the other for total development.

Once again, we are poised for a bitter battle where emotions are high, facts are few, and a number of important national issues are at stake.

What happened? Why do we find ourselves at the edge of a "black hole" of public policy, asked to take a leap of faith into the unknown? Sadly, the answer is simple: In the "what you don't know won't hurt you" theory of government that has dominated Washington, D.C. in recent years, political ideology prevails over knowledge.

Some people—including those in recent administrations—clad the need for oil and gas development in the patriotic cloak of "national security." The Reagan Administration, for example, rushed the nation into a decision about oil field development, in spite of knowing very little about the possible impacts on one of the world's most sensitive ecosystems.

Questions That Must Be Answered

We have not answered questions that must be answered before a Congressional decision can be made about opening the Coastal Plain of the Arctic National Wildlife Refuge to oil development or maintaining its current protected status. For example, do we really know enough about the potential oil reserves of the 1002 area? No. And I believe we must know what exists there even if we decide that the nation's best strategic course requires deferral of extraction for another fifty years.

Other questions linger. For example, do we really understand the probable impacts of development on the internationally invaluable porcupine caribou herd or on the area's musk oxen population or other fish and wildlife resources? *No.* Do we know the environmental impacts of full oil-field development on the area's air and water quality or the effects of toxic substance bioaccumulation? *No.* Has anyone evaluated the cumulative impacts of circumpolar development on the arctic environment and its wild living resources? *No.*

I could go on, but I think I've made my point. A coordinated, long-term research program was not undertaken before the critical question of oil and gas development in this arctic ecosystem was presented.

The Arctic National Wildlife Refuge is not the only such instance. We continue to make the same kinds of mistakes on a wide range of important public policy issues. Our society must learn that in order to make responsible decisions among competing and complex choices, all interests will be best served if better science and enhanced information transfer become more integral elements of the public decision-making process.

Surely, if we can commit billions of research dollars to the development of a dubious space-based defense program, then we should commit millions of dollars to environmental research designed to understand the life-support systems of this planet of which we are but one part.

No Better Gift

In the very early 1960s, President John Kennedy pledged to put a man on the moon. In 1969, we accomplished that feat. Wouldn't it be just as worthy of a president today to commit our nation to a comprehensive inventory of the world's wild living resources by the year 2000? Aside from a world at peace with itself, I know of no better gift we could leave to the children of the twenty-first century.

Let me make a couple of final points about the process of scientific research.

One of the most important lessons I learned about scientific research came during my graduate school days at the University of Alberta in Canada. I had just presented to my major professor the data from my doctoral dissertation on the quantification of the structures and function of a complex biological community.

Without a word, he looked carefully through my reams of computer printouts and graphs. After an hour or more, he looked up and said, "These are the most incredible answers I have ever seen. . . . Do you have any idea what the questions are?"

Whether by design or chance, his response sent a tidal wave of fear through me. Fortunately, after I regained my composure, I convinced him that I did, indeed, have some idea of what the questions were.

The point he made so succinctly has remained with me: Scientific research is conducted within a framework of developing and testing hypotheses. That lesson must apply as we try to answer scientific questions relating to wilderness ecosystems.

Frankly, we need to generate and test more rigorous hypotheses at every stage of the research process. Likewise, we need to reallocate our research priorities and our research dollars. Haven't we counted enough elk feces? Do we really need the ten-thousand-and-first research project on the white-tailed deer when the species is flourishing, and at least 10 million dollars have been spent on research since 1950?

Wouldn't it be more valuable to fund long-term research programs into such questions as: How do wilderness ecosystems function? What species are present? What is their relative stability over time? What variables are most critical? What happens when they are perturbed by natural causes? By human activities? What resource management knowledge can we apply to nonwilderness areas?

And wouldn't it be more valuable to quantify the demand curves for wilderness recreation or its contribution to our Gross National Product? Wouldn't it be more valuable to assess our land management policies regarding *all* public lands in order to determine how much acreage should be designated as "wilderness"? Wouldn't it be more valuable to evaluate how many miles of roads we can build in our national forests before we end up with a highway system separated by strings of trees and silted streams instead of an integrated forest ecosystem capable of sustaining a broad array of renewable natural resources?

And, finally, given the scale of the worldwide destruction of wilderness ecosystems and the limited financial resources at our disposal, wouldn't it be prudent to systematically identify those areas of greatest importance and aggressively proceed to protect them? This priority-ranking approach, sometimes known in medical circles as a "triage strategy," would not be without controversy. Who decides, for example, what areas are most important?

However, as Norman Myers (1986) noted, far from seeking to establish quantification of all critical parameters, a triage approach tries to identify all relevant sets of values in order to illuminate an unduly confused situation. Such an approach would bring a degree of order to the current haphazard process and allow us to make the best use of available financial and other resources. By emphasizing the protection of entire communities of species or entire ecosystems, we could avoid the moral dilemmas inherent in a triage approach as it relates to saving individual endangered species.

In short, we need more emphasis on the importance of natural resources–related scientific and socioeconomic research to meet the needs of modern society. And we need to approach such research more creatively, even if it generates some controversy.

A New Attitude

In addition to a new research direction, we need a new attitude. First, it is important to remember that science is only orderly after the fact. During the research process—and particularly on the frontiers of research—science can be chaotic and fiercely controversial. Likewise, we need to be more cautious in characterizing research as either "basic" or "applied." While there may be some truth in the definition that a "specialist" is someone who "knows more and more about less and less," there is another side to that coin. A tremendous idea in science often appears to have its birth as a particular answer to a narrow question. Many times, it is much later when the ramifications of that answer become apparent. What began as knowledge about very little often turns out to be wisdom about a great deal.

As Louis Pasteur said, "There is no such thing as applied science. There are only applications of science."

Second, we must bury the adversarial relationships that have existed too long among various sectors of our society. Isn't it time that the timber industry and conservation interests stop drawing battle lines and start charting an effective and positive strategy for both economic development and enhancement of the wilderness system? Can't we agree that if we bring better information and less rhetoric to the decision-making process, we will produce better public policies?

Finally, we must do a far better job of moving new information into the public policy and resource management arenas. Relevant research must reach the table where decisions are made.

From the applications of science will come understanding. And from understanding will come new and creative opportunities for meeting the needs of society. Acting on those opportunities will present a challenge to all of us. To meet those challenges, we need leaders who can set aside narrow, provincial thinking and adopt the broader goal of a nation secure in both its economic vitality and in the conservation of its natural resources. A nation whose people, while first and foremost Americans, are also citizens of the world.

We need leaders who are willing to take risks, but not with the health of our environment or the natural heritage we hold in trust for future generations. We need leaders to educate our society and provide the scientific knowledge for continued advancement. We need leaders from all walks of life who have inspired visions of a better tomorrow and a sense of stewardship for those yet unborn.

Finally, as we face the leadership challenges, we should take to heart the words of a great conservationist, President Theodore Roosevelt: "Far better it is to dare mighty things, to win glorious triumphs, even though checkered by failure, than to take rank with those who neither enjoy much nor suffer much because they live in the gray twilight that knows not victory nor defeat."

We must resolve not to live in that gray twilight but, rather, to search the vastness of wild places for that which we have lost—and for that which we have not yet found. We must "hear the challenge, learn the lessons, pay the costs." For ours, among all generations, is literally being given the last chance to save the best of that which remains of our wilderness heritage. We dare not fail our duty.

References

Myers, N. Tackling mass extinction of species: A great creative challenge.
Albright Lecture. Berkeley: University of California; 1986.

Raven, P. H. We're killing our world. Keynote Address at the American
Association for the Advancement of Science Meeting, Chicago; 1986.

Service, R. W. "The call of the wild." In: Carman, B., ed. Canadian poetry
in English. Toronto: Ryeson Press; 1954.

Wilson, E. O. The biological diversity crisis: A challenge to science. Issues
in Science and Technology 2:20–29; 1985.

A Distant Perspective on
the Future of Americans Outdoors

L. DAVID MECH

T H E R E ' S nothing like a clean arctic breeze, a high midnight sun, and a barren, mountainous vista stretching toward the end of the earth to give one perspective. My presence not long ago in such a spot, just a few hundred miles from the north pole, amidst musk-ox and wolf, Peary caribou and arctic hare, was truly the highlight of a career of outdoor work.

And an appropriate place it was for pondering the plight of the planet and reflecting about the future of the outdoors. My surroundings seemed ageless. Except for an occasional high-flying jet scribing a lonely polar route, the scene probably was the same 100 or 1,000 years before. Not true, the world to the south. It had changed rapidly, not always for the better.

I thought about my urban home in the United States. Not only was it thousands of miles away, but thousands of years as well. Those years had brought a human population with a culture and philosophy and, most important of all, a technology, that could transform any natural piece of the earth to a grotesque tangle of artificiality. But to a comfortable artificiality—more to the liking of the modern human. So the species became fruitful and multiplied. And multiplied. And multiplied again. And between a rampantly increasing populace and a technology with a life of its own, a whole new world had been created. Instead of terra firma, there was concrete and asphalt. Where tall trees once grew, brick and metal buildings sprouted. Streams became dirty ditches; ponds and lakes, cesspools. And the air fouled the lungs of the old and sick. Even the rain bore the destructive by-products of this new force.

But I realized that that is the real world of most of the country's current population. Their natural habitat is mortar, steel, and glass. Their wildlife consists of dogs, cats, and pigeons. Their milk originates in plastic jugs, and their meat in tightly wrapped styrofoam packages with parsley sprigs. Although most people realize that a natural world exists somewhere, their knowledge of it is necessarily artificial, superficial, and emotional. Their view of it is two-dimensional with no smell or feel, but with plenty of hearts, flowers, and violins. This world fits on a nineteen-inch screen and can be turned on and off at will. These people are mere spectators to nature. Their numbers are growing rapidly.

Certainly some people occasionally, or even regularly, slip their bonds to artificiality and actually experience the natural world. However, as a proportion of the total population, they represent a small fraction. Even when they do rough it, most do so from their auto, recreational vehicle, or camper. Others commune with nature from power boats, snowmobiles, or all-terrain vehicles. These conveyances are becoming increasingly sophisticated, and more and more people are using them.

This leaves a portion of the population who hike, ski, canoe, sail, climb, snowshoe, kayak, hang-glide, camp, or photograph. These non-consumptive users experience the outdoors much more intimately. Their numbers are relatively few, but they too are increasing.

The outdoors users who actually participate in the natural world, however, are decreasing in proportion. They are the consumptive users of the outdoors: primarily fishermen, hunters, and trappers, but also mushroom hunters, berry-pickers, and such. These people obtain their enjoyment of nature by interacting with its animals or plants at a level decidedly more intense than that of most other outdoor lovers. However, many of their traditional practices are now coming under attack by people who fail to understand their viewpoint of the natural world.

Thus, as I reflected on all the varied uses of the outdoors by an increasingly urbanized populace, several disturbing trends became apparent when I speculated about the future. First and foremost is the tremendous pressure generated by our burgeoning population. Such pressure promotes the continued destruction of what is left of nature, its soil, air, water, vegetation, and wildlife, and furthers the development of an ever more artificial world. This pressure also directly threatens the quality and quantity of outdoor experience available. In many parts of the wilderness these days, one must stand in line to portage a canoe. Reservations for campsites are increasingly a reality.

Secondly, I saw a more distorted view of the natural world being fostered by the increasingly urbanized living and the dissemination of information about nature primarily via television. Although television's promotion of an interest in nature is valuable, the superficiality of the information worries me greatly. People's knowledge about nature has become two-dimensional. Once while conducting an experiment in a zoo, I had placed a pair of wolves in an empty leopard cage. Because the sign on the cage still read "leopard," many passing people informed their children authoritatively that the wolves were leopards.

Such misinformation may seem harmless. However, it can lead to absurdities like the question I have twice been asked as a wolf biologist: "Why can't the government catch all the old, sick, and weak deer and moose, euthanize them, and let the wolves feed on them? Then the wolves wouldn't have to kill the animals." The people who asked the question couldn't bear the thought of carnivores killing "innocent" prey, although this process is a constant, vital part of nature.

This leads me to the third disturbing trend I saw relevant to human use of the outdoors,—the increasing dissension about such use. The same attitude that views predation as evil, I believe, drives the current movement against hunting, trapping, and fishing. Whereas, science regards human beings as having evolved as predators, much of the public now considers humans as outside of nature rather than as a part of it. This view fosters a strong protectionism: "Save the whale," "Stop killing baby seals," "Furs looks best on their original owners."

While I've never thought it particularly nice to kill a baby seal, I also know that the situations these slogans address are not quite as simple as implied. It certainly is true that some species need strong protection. However, the protectionist attitude has been applied to deer, geese, muskrats, and various other species that often do considerable damage or, at the very least, occur in high numbers. A person's desire not to kill a creature of any sort, of course, must be respected. The problem comes when the person tries to impose his/her ethical view on everyone else.

Dissension about use of the outdoors is broader than just the anti-harvesting controversy, however. Skiers complain about snowmobilers; canoeists about motorboaters; and hikers about all-terrain vehicles. Fly-fishermen look down on those who use live bait. In some states, fox hunters have even forced through legislation preventing fox trapping.

The last trend that concerns me is the application of technology to outdoor gear and equipment. This trend is really a mixed blessing, I believe. Certainly, it is wonderful that better boots and clothing now allow more people to enjoy the outdoors during all seasons and in all kinds of weather. However, it does mean additional pressure on outdoor resources. Furthermore, this same technology has now blessed us with pickup trucks bearing tires so large their only real use can be to destroy things. Such vehicles do allow grown men to enjoy getting stuck in mud holes, I admit. But what does that do to the environment? Various types of all-terrain vehicles actually bother me more, not only because of the environmental destruction they themselves can wreak, but also because of the additional accessibility, and therefore pressure, they allow to areas previously untouched.

What all this adds up to is a future in which an increased population with a decreased knowledge of the natural world will have the wherewithal to use the outdoors to a degree now unimagined. And the potential to destroy it to the same extent.

What to do? There are several things, I believe, that will help.

Population control is certainly one of the most efficient and effective means of alleviating problems at least partly caused by population pressure. It seems inevitable that any sane and rational society will eventually try to regulate its numbers. We have a ways to go in that respect. But the stakes are high, not only in terms of the quantity and quality of outdoor recreation in the future, but also in social, ecological, economic, and aesthetic terms. I have faith that our society will eventually grasp this truth. However, we had better at least start publicly talking about it soon.

A second worthy approach is to preserve, as soon as possible, as much wild and natural land and habitat as we can. We can never manufacture any more. We already have a good start in our national parks and wilderness systems, wildlife refuges, research natural areas, wild and scenic rivers, and nature centers, and the high use of these areas attests to their value. Because of the limited nature of the earth's remaining natural areas, however, they will forever increase in value. Thus, I cannot stress too strongly the extreme need to continue to preserve such areas, and to acquire and preserve more of them, large or small. To a child growing up in a city or suburb, a vacant lot can be a valuable wilderness; too bad there aren't more.

Although acquisition and preservation of extensive natural areas should continue to be a function of the federal government because of its large land holdings, local governments should also play an increased role. Some municipalities already require that a certain percentage of each newly developed parcel be set aside as "open space." That trend should be repeated everywhere, with county and state governments following suit. In this respect, the work of The Nature Conservancy bears mention. That organization seeks out unique natural areas, buys them to preserve them, and then turns them over to appropriate government agencies to administer and protect forever. Except for Planned Parenthood, I regard The Nature Conservancy as the most effective, most efficient conservation organization there is.

Here it is necessary to distinguish between the preservation of natural areas and the preservation of wildlife. Natural areas are not renewable. Destroy them and they are gone. Wildlife, however, is renewable. So long as the taking of wildlife is regulated, as it now is almost everywhere, one need not worry about harvesting it. (Endangered species are exceptions, but most of them were reduced by factors other than harvesting.)

It is important to understand this distinction between renewable and nonrenewable outdoor resources. I see the need to preserve natural areas as so important that every preservation dollar spent trying to "save" renewable wildlife instead of its habitat seems counterproductive. Certainly it is easier to persuade people to donate money, time, or effort to prevent the killing of some cuddly animal. However, what good does it do to outlaw the trapping of muskrats, for instance, when someone is draining the marsh in which they live? Save the marsh, however, and the population of muskrats, mink, raccoons, turtles, frogs, fish, and numerous birds will take care of themselves.

The only real exception involves endangered species. They too need habitat, of course, and for some of them, that is the major problem. For others, however, deliberate efforts to study them, protect them, breed them in captivity, or reestablish them in the wild are necessary. A prime example is the restoration of the wolf to Yellowstone National Park. Because wolves preyed on livestock, they were deliberately wiped out of the extensive western wilderness, including national parks where there were no livestock. This left Yellowstone an incomplete ecosystem. Restoring the wolf to that park is biologically feasible and necessary, and it would greatly increase the enjoyment of Yellowstone by its millions of visitors.

Another major need I see for helping solve the problem of increasing use of the outdoors in the future is that of maintaining as much of a pluralistic approach as possible. Preservation of natural areas need not mean prohibiting human use; it only means prohibiting destructive use. True, some outdoor uses may be mutually exclusive when practiced in the same location at the same time. Canoes and powerboats are incompatible. So are fox hunting and fox trapping. However, these problems should be remedied by zoning, rather than by complete outlawing of one or the other. For example, once pristine wildernesses are protected from such destructive use as brought by all-terrain vehicles (ATVs); other areas, already destroyed, can be set aside for ATVs. Perhaps we even need national mud holes specially designated for people who like four-wheel driving!

With the proper seasonal zoning, much greater use could also be made of any given area. For example, public open space situated near cities but isolated enough from residences could serve for hiking, picnicking, and even camping during summer, and then for hunting and trapping in fall and winter. Fishing could be allowed year-round. Not only would this approach provide more people with more outdoor opportunities, but it would help take pressure off other areas. Of course, this presupposes that there will be enough public open spaces and natural areas acquired and set aside to begin with.

How can all these needs be taken care of, forming a national population conscience, preserving as many natural areas as possible, encouraging distinctions between renewable and nonrenewable outdoor resources, restoring endangered species, and promoting enlightened and ecologically sound pluralism in use of the outdoors?

By outdoor education, I hope. But here is where we have really fallen down. High school and college courses in biology, ecology, conservation, and wildlife management are especially helpful, and the increasing contributions of museums and nature centers are also valuable. However, far more needs to be done.

The outdoor education challenge must be met head on. Ecology and conservation courses need to be instituted in every grade school and high school, and in extension, adult special, and night classes. More nature centers are necessary, and more nature classes should be offered by museums, zoological societies, and conservation organizations. A strong concerted effort is needed.

However, the real leadership in outdoor education should come from the state departments of natural resources and the federal government. Most states publish a conservation magazine and offer outdoor movies, and that is good so far as it goes. The federal agencies charged with managing land, forests, parks, and wildlife also include nominal information and education (I & E) divisions. However, most often such divisions operate much the same as they did 30 years ago. And when budget-cutting time comes, these are often the first divisions to feel the axe.

The popularity of television and home video machines has brought significant opportunities for I & E divisions that have yet to be exploited. Home video equipment should be issued to every project leader, and the leader trained, so that opportunity footage can be obtained of their work. I & E divisions themselves should have adequate professional video equipment, and access to editing facilities, so they can produce their own videos.

Most of all, conservationists in leadership positions, both public and private, must concentrate their energies on educating the media about the above problems and their solutions. We have been saturated with superficiality; the public may even find relief in exposure to subjects in greater depth.

But the time for all this is now. Every day that passes sees more pavement laid, more marshes drained, and more basements dug. From my vantage point in the high arctic, where wolves still chase the musk-ox in a world unknown to most, the future to the south seems much too imminent.

Bibliography of Wilderness Books

Following is a listing of selected books on wilderness and related themes. This is but a relatively short list on the topic; interested readers should consult libraries, bookstores, and the reference lists in this volume's chapters for other titles.

Abbey, E. Desert solitaire: A season in the wilderness. New York: Touchstone Books, Simon and Schuster; 1968.

Agee, J. K.; Johnson, D. R., editors. Ecosystem management for parks and wilderness. Seattle: University of Washington Press; 1988.

Allin, C. W. The politics of wilderness preservation. Westport, CT: Greenwood Publishing Group; 1982.

Auerbach, P. S.; Geehr, E. C., editors. Management of wilderness and environmental emergencies. New York: Macmillan; 1983.

Banks, V. The Pantanal: Brazil's forgotten wilderness. San Francisco: Sierra Club Books; 1991.

Bonney, O. H. Guide to the Wyoming mountains and wilderness areas. Athens: Ohio University Press; 1977.

Brooks, P. The pursuit of wilderness. Boston: Houghton Mifflin; 1971.

Brower, D. R., editor. Wildlands in our civilization. San Francisco: Sierra Club Books; 1964.

———. For Earth's sake: The life and times of David Brower. Salt Lake City, UT: Peregrine Smith Books; 1990.

———. Work in progress. Salt Lake City, UT: Peregrine Smith Books; 1991.

Carroll, P. N. Puritanism and the wilderness: The intellectual significance of the New England frontier. New York: Columbia University Press; 1969.

Cicchetti, C. J.; Smith, V. K., editors. The costs of congestion: An econometric analysis of wilderness recreation. Cambridge, MA: Ballinger Publishing Company; 1976.

Crowley, K.; Link, M. Boundary waters canoe area wilderness. Stillwater, MN: Voyageur Press; 1987.

Crump, D. J., editor. America's hidden wilderness: Lands of seclusion. Washington, DC: National Geographic Society; 1988.

Dasmann, R. F. A different kind of country. New York: Macmillan; 1968.

Douglas, D. Wilderness sojourn: Notes in the desert silence. San Francisco: Harper & Row; 1987.

Elbers, J. S., compiler. Changing wilderness values, 1930–1990: An annotated bibliography. Westport, CT: Greenwood Publishing Group; 1991.

Fisher, J. L. Notes on the value of research on the wilderness part of wildland. Washington, DC: Resources for the Future; 1960.

Foreman, D.; Wolke, H. The big outside: A descriptive inventory of the big wilderness areas of the United States. Tucson, AZ: Ned Ludd Books; 1989.

Fradkin, P. L. Sagebrush country: Land and the American West. New York: Knopf; 1989.

Frome, M. Battle for the wilderness. Boulder, CO: Westview Press; 1974.

———. Issues in wilderness management. Boulder, CO: Westview Press; 1984.

Gillette, E. R. Action for wilderness. San Francisco: Sierra Club Books; 1979.

Glover, J. M. A wilderness original: The life of Bob Marshall. Seattle, WA: The Mountaineers; 1986.

Graf, W. L. Wilderness preservation and the sagebrush rebellions. Savage, MD: Rowman & Littlefield; 1990.

Haines, J. M. The stars, the snow, the fire: Twenty–five years in the northern wilderness. St. Paul, MN: Graywolf Press; 1989.

Hammit, W. E.; Cole, D. N. Wildland recreation: Ecology and management. New York: John Wiley & Sons; 1987.

Hendee, J. C.; Stankey, G. H.; Lucas, R. C. Wilderness management. Washington, DC: U.S. Dept. of Agriculture Forest Service Miscellaneous Publication Number 1365; 1978.

Irland, L. C. Wilderness economics and policy. Lexington, MA: Lexington Books; 1979.

Keiter, R. B.; Boyce, M. S., editors. The greater Yellowstone ecosystem: Redefining America's wilderness heritage. New Haven, CT: Yale University Press; 1991.

Kilgore, B. M., editor. Wilderness in a changing world. 9th Wilderness Conference (1965), Sierra Club. San Francisco: Sierra Club Books; 1966.

LaBastille, A. Women and wilderness. San Francisco: Sierra Club Books; 1984.

————. Beyond Black Bear Lake: Life at the edge of wilderness. New York: Norton; 1988.

Leopold, A. A Sand County almanac and sketches here and there. New York: Oxford University Press; 1949.

Littlejohn, B. M.; Pimlott, D. H., editors. Why wilderness? A report on mismanagement in Lake Superior Provincial Park. Toronto: New Press; 1971.

Lucas, R. C., compiler. Proceedings—National Wilderness Research Conference: Current research. Ogden, UT: U.S. Dept. of Agriculture Forest Service General Technical Report INT-212; 1986.

————. Proceedings—National Wilderness Research Conference: Issues, state-of-knowledge, future directions. Ogden, UT: U.S. Dept. of Agriculture Forest Service General Technical Report INT-220; 1987.

Lyon, T. J., editor. This incomperable lande: A book of American nature writing. Boston: Houghton Mifflin; 1989.

Martin, V.; Inglis, M., editors. Wilderness, the way ahead. Forres, Scotland: Findhorn Press; 1984.

Matthiessen, P. The cloud forest: A chronicle of the South American wilderness. New York: Viking Penguin: 1987.

McCloskey, M. E., editor. Wilderness; The edge of knowledge. 11th Wilderness Conference (1969), Sierra Club. San Francisco: Sierra Club Books; 1970.

McCloskey, M. E., editor. Wilderness: The edge of knowledge. 11th Wilderness Conference (1969), Sierra Club. San Francisco: Sierra Club Books; Club Books; 1969.

McKibben, B. The end of nature. New York: Random House; 1989.

McPhee, J. A. Encounters with the archdruid. New York: Farrar, Straus and Giroux; 1971.

Miller, D. S. Midnight wilderness: Journeys in Alaska's Arctic National Wildlife Refuge. San Francisco: Sierra Club Books; 1990.

Miller, M.; Wayburn, P. Alaska, the great land. San Francisco: Sierra Club Books; 1974.

Muir, J. Wilderness essays. Salt Lake City, UT: Peregrine Smith Books; 1980.

Nash, R. Wilderness and the American mind. 3d ed. New Haven, CT: Yale University Press; 1982.

National Geographic Society Staff, editors. Wilderness U.S.A. Washington, DC: National Geographic Society; 1975.

Northern Lights Staff, editors. The wilderness gridlock. Missoula, MT: Northern Lights Research & Education Institute; Fall issue; 1991.

Oelschlaeger, M. The idea of wilderness: Prehistory to the age of ecology. New Haven, CT: Yale University Press; 1991.

―――, editor. The wilderness condition: Essays on environment and civilization. San Francisco: Sierra Club Books; 1991.

Peck, R. M. Land of the eagle, a natural history of North America. New York: Summit Books; 1991.

Petzoldt, P. The wilderness handbook. New York: W. W. Norton & Company; 1984.

Rasmussen, A., editor. Wilderness areas: Their impacts. Proceedings of a symposium. Logan: Cooperative Extension Service, Utah State University; 1990.

Ronald, A., editor. Words for the wild. San Francisco: Sierra Club Books; 1987.

Schoenfeld, C.; Hendee, J. C. Wildlife management in wilderness. Pacific Grove, CA: Boxwood Press; 1978.

Schullery, P., editor. Theodore Roosevelt: Wilderness writings. Salt Lake City, UT: Peregrine Smith Books; 1986.

Schwartz, W., compiler. Voices for the wilderness. New York: Ballantine Books; 1969.

Sholly, D. R.; Newman, S. M. Guardians of Yellowstone: An intimate look at the challenges of protecting America's foremost wilderness park. New York: Morrow; 1991.

Simer, P.; Sullivan, J. The National Outdoor Leadership School's official wilderness guide. Austin, TX: S&S Press, 1983.

Spears, B., editor. Wilderness Canada. Toronto: Clarke, Irwin and Company; 1970.

Stegner, W. The American West as living space. Ann Arbor: University of Michigan Press; 1987.

Teale, E. W., editor. The wilderness world of John Muir. Boston: Houghton Mifflin; 1954.

Trimble, S. The sagebrush ocean: A natural history of the Great Basin. Reno: University of Nevada Press; 1989.

Turner, T. Wild by law: The Sierra Club Legal Defense Fund and the places it has saved. San Francisco: Sierra Club Books; 1989.

United States Bureau of Land Management. Wilderness inventory handbook: Policy, direction, procedures, and guidance for conducting wilderness inventory on the public lands. Washington, DC: Department of the Interior, Bureau of Land Management; 1978.

Vickery, J. D. Wilderness visionaries. Merrellville, IN: ICS Books; 1988.

Wertheim, A. The intertidal wilderness. San Francisco: Sierra Club Books; 1984.

Wheelwright, J. H.; Schmidt, L. W. The long shore: A psychological experience of the wilderness. San Francisco: Sierra Club Books; 1990.

Wilderness Society Staff, editors. Wilderness America: A vision for the future of the nation's wildlands. Washington, DC: Wilderness Society; 1989. Distributed by: Gibbs Smith, Salt Lake City, UT.

Zaslowsky, D. These American lands: Parks, wilderness, and the public lands. New York: H. Holt; 1986.

Zeveloff, S. I.; McKell, C. M., editors. Wilderness issues in the arid lands of the western United States. Albuquerque: University of New Mexico Press; 1992.

Contributors

Richard M. Alston
Department of Economics
Weber State University
Ogden, Utah 84408-3807

Anne Berman
76 Remsen Street
New York, New York 11201

Chris Bullock
Department of English
University of Alberta
Edmonton, Alberta
Canada T6G 2E5

Russell Burrows
Department of English
Northwest Wyoming College
Powell, Wyoming 82435

Felicia F. Campbell
Department of English
University of Nevada, Las Vegas
Las Vegas, Nevada 89154

Cheryl Charles
Project WILD
P.O. Box 18060
Boulder, Colorado 80308-8060

Michael P. Cohen
Department of English
Southern Utah University
Cedar City, Utah 84720

Terrell Dixon
Department of English
University of Houston
Houston, Texas 77000

Thomas L. Fleischner
Department of Environmental
Studies
Prescott College
Prescott, Arizona 86301

Jay D. Hair
National Wildlife Federation
1400 Sixteenth St., N.W.
Washington, D.C. 20036-2266

William C. Johnson, Jr.
Division of Literature and
Languages
Lewis-Clark State College
Lewiston, Idaho 83501-2698

William Kittredge
Department of English
University of Montana
Missoula, Montana 59801

Howard McCord
Department of English
Bowling Green State University
Bowling Green, Ohio 43402

William H. McVaugh
Department of Psychology
Weber State University
Ogden, Utah 84408-1202

L. David Mech
U.S. Fish and Wildlife Service
North Central Forest Experiment
Station
1992 Folwell Ave.
St. Paul, Minnesota 55108

George Newton
Alberta Environmental Network
10511 Saskatchewan Drive
Edmonton, Alberta
Canada T6E 4S1

Cliff Nowell
Department of Economics
Weber State University
Ogden, Utah 84408-3807

Wayne Ouderkirk
Empire State College
Cobleskill, New York 12043-1701

Ann Ronald
College of Arts and Science
University of Nevada, Reno
Reno, Nevada 89557-0008

Neila C. Seshachari
Department of English
Weber State University
Ogden, Utah 84408-1201

Steven R. Simms
Department of Sociology, Social
Work, and Anthropology
Utah State University
Logan, Utah 84322-0730

L. Mikel Vause
Department of English
Weber State University
Ogden, Utah 84408-1201

Edward B. Weil
School of Liberal Arts
Kean College of New Jersey
Union, New Jersey 07083

Brooke Williams
3499 Little Tree Road
Salt Lake City, Utah 84108

Terry Tempest Williams
3499 Little Tree Road
Salt Lake City, Utah 84108

Samuel I. Zeveloff
Department of Zoology
Weber State University
Ogden, Utah 84408-2505

Authors' Biographies

RICHARD M. ALSTON, professor and chair of the Department of Economics at Weber State University where he also holds a Willard L. Eccles Research Fellowship in Natural Resource Policy Analysis, has published articles in the *Journal of Forestry*, *Environmental Law*, *Ecology Law Review*, and the *Journal of Environmental Management* among others.

ANNE BERMAN, formerly a landscape architect with the City of New York Department of Parks and Recreation and with private landscape firms, is a consultant to residential and commercial design projects.

CHRIS BULLOCK, professor of English at the University of Alberta in Edmonton, Alberta, Canada, is the joint author of *A Guide to Marxist Literary Criticism* (1980) and *Essay Writing for Canadian Students* (1985) and has written articles on modern literature, the environment, nature writing, and the men's movement.

RUSSELL BURROWS, assistant professor of English at Northwest College of Wyoming, has contributed articles to *Western American Literature*, *Petroglyph: A Journal of Creative Natural History Writing*, *Dialogue: A Journal of Mormon Thought*, and *Rocky Mountain Gardener*. He is also a staff bibliographer for the *MLA International Bibliography of Books and Articles on the Modern Languages and Literatures*.

FELICIA CAMPBELL, associate professor of English at the University of Nevada, Las Vegas, has contributed to, among others, *Time, Rhythms and Chaos in the New Dialogue with Nature* and *Studies in Frank Waters*. She is Executive Director of the Far West Popular Culture Association and editor of *Popular Culture Review*.

CHERYL CHARLES, Director of Project WILD, an award-winning environment education program for K-12 teachers, is the author of *The WholeSchool Book: Teaching and Learning in the Late Twentieth Century* and *Report and Proceedings: Snowmass Summit on Education, Wildlife and the Environment*. Her awards include the 1991 Professional of the Year Award from the Western Association of Fish and Wildlife Agencies.

MICHAEL PETER COHEN, professor of English at Southern Utah University, is the author of *The Pathless Way: John Muir and American Wilderness* and *The History of the Sierra Club*.

TERRELL F. DIXON, chair of the English Department at the University of Houston, Central Campus, has written "Edward Abbey's Biocentric Epiphany: *Desert Solitaire* and the Teaching of Environmental Literature" for a recent CEA Critic special issue of nature writing, is co-editor of *Being in the World: An Environmental Reader for Writers* to be published by Macmillan, and is editing a collection of essays on Edward Abbey and an interdisciplinary anthology of the grizzly bear.

THOMAS L. FLEISCHNER, professor of Environmental Studies at Prescott College in Prescott, Arizona, and co-founder and Program Director of the North Cascades Institute, is the author of articles dealing with biological diversity and conservation education. He has been involved in ecological field research on marine birds and mammals from Puget Sound to Alaska and from Maine to Mexico and with resource and wilderness management of the Greater North Cascades Ecosystem, the Colorado Plateau, and the Sonoran Desert.

JAY D. HAIR, president and chief executive officer of the National Wildlife Federation and chairperson of the Board of Directors and president of the IUCN-US (The World Conservation Union), has written extensively on environmental issues. Recent awards include Environmental Educator of the Year (Ball State University, 1988) and the 1990 National Park Foundation Theodore and Conrad Wirth Environmental Award.

WILLIAM C. JOHNSON, JR., professor of English at Lewis-Clark State College, in Lewiston, Idaho, is the author of various publications, including *What Thoreau Said: 'Walden' and the Unsayable*, and is currently completing a volume of poetry, *At the Wilderness Boundary*.

WILLIAM KITTREDGE, professor of English at the University of Montana, has written stories and essays for a number of magazines including *Outside, Harper's*, and *Rolling Stone*, and has also published a collection of short stories, *We Are Not in This Together* and a collection of essays, *Owning It All*.

HOWARD MCCORD, professor of English at Bowling Green State University and the author of more than twenty-five books, is currently writing a book about the Jornada del Muerto, a ninety-mile stretch of waterless desert in New Mexico.

WILLIAM H. MCVAUGH, professor of psychology at Weber State University and formerly an exploration geologist, has engaged in research in human cognition and clinical psychology and worked with children and adolescents in a variety of clinical settings.

L. DAVID MECH, wildlife research biologist with the U.S. Fish and Wildlife Service and adjunct professor in Ecology and Behavioral Biology as well as Fisheries and Wildlife at the University of Minnesota, is the author of various publications, including *Wolves of Isle Royale, The Wolf: Ecology and Behavior of an Endangered Species*, and *The Arctic Wolf*. His awards include two Special Achievement Awards from the U.S. Fish and Wildlife Service (1971, 1981) and a Gulf Oil Professional Conservationist Award (1984).

GEORGE NEWTON, provincial coordinator for the Alberta Environmental Network and chair of the Edmonton chapter of the Canadian Parks and Wilderness Society, is the author of a thesis on the significance of river experiences for environmental education.

CLIFF NOWELL, associate professor of economics at Weber State University where he also holds a Willard L. Eccles Research Fellowship in Environmental Economics, has published widely in such journals as *Ecological Economics, Land Economics, The Southern Economics Journal*, and *The Review of Economics and Statistics*. In addition to environmental economics, Nowell's research includes industrial organization, research methodology, and public utility regulation.

WAYNE OUDERKIRK, faculty member at Empire State College, the non-traditional college of the State University of New York, is currently at work on several philosophy articles defending the notion of natural intrinsic value as a basis for environmental ethics.

ANN RONALD, dean of the College of Arts and Science and professor of English at the University of Nevada, Reno, is the author of *The New West of Edward Abbey*, editor of *Words for the Wild*, and is currently working on a book about the Great Basin in Nevada.

NEILA C. SESHACHARI, professor of English and editor of *Weber Studies: An Interdisciplinary Humanities Journal* at Weber State University, has published a number of articles in professional journals and books. Her research interests include women's studies, Asian and immigrant literatures, and contemporary American literature.

STEVEN R. SIMMS, associate professor of anthropology at Utah State University, has written about Great Basin archaeology, the evolutionary ecology of hunter-gatherers, the ethnoarchaeology of Bedouin in the Near East, and archaeological method and theory.

L. MIKEL VAUSE, associate professor of English at Weber State University, is editor of *Rocks & Roses: A Collection of Essays by Women Mountaineers, Rocks & Roses, Volume II*, and a mountaineering anthology to be published by Gibbs Smith. He has also published numerous articles on mountaineering, wilderness literature, and literary criticism.

EDWARD B. WEIL, dean of the School of Liberal Arts and professor of anthropology at Kean College of New Jersey, was principal investigator on numerous archaeological projects in California and directed the cultural resource components of several major environmental impact studies involving public lands in California and the U.S. Southwest.

BROOKE WILLIAMS is a freelance journalist, business consultant, and wilderness traveler.

TERRY TEMPEST WILLIAMS, naturalist-in-residence at the University of Utah's Museum of Natural History, is the author of *Pieces of*

White Shell: A Journey to Navajo Land which received a Southwest Book award, *Coyote's Canyon*, two children's books, and *Refuge: An Unnatural History of Family and Place*. She serves on the Governor's Council of The Wildlife Society in Washington, D.C.

SAMUEL I. ZEVELOFF, chair and professor in the Department of Zoology at Weber State University, is the author of *Mammals of the Intermountain West*, and *Wilderness Issues in the Arid Lands of the Western United States* (edited with C.M. McKell), and articles on mammalian evolutionary ecology in *The American Naturalist*, *Evolution*, and *Nature*.

Credits

Burrows, R. W. Stegner's version of pastoral. Western American Literature. 25:15–25; 1990. Copyright 1990 by Western American Literature. Reprinted by permission of Western American Literature.

Hair, J. D. Wilderness: Promises, poetry, and pragmatism. In: University of Idaho Wilderness Resource Distinguished Lectureship Series. Moscow, ID: University of Idaho Wilderness Research Center; 1987. Copyright 1987 by University of Idaho, Wilderness Research Center. Reprinted by permission of the University of Idaho College of Forestry, Wildlife, and Range Sciences.

Kittredge, W. Yellowstone in winter. In: Kittredge, W. Owning it all. St. Paul, MN: Graywolf; 1987. Copyright 1987 by William Kittredge. Reprinted by permission of Graywolf Press.

McCord, H. The arctic desert. In: McCord, H. Walking edges. Ohio: Raincrow Press; 1982. Copyright 1982 by Howard McCord. Reprinted by permission of the author.

Mech, D. L. A distant perspective on the future of American outdoors. In: Philosophical essays on recreation and the outdoors. Washington, DC: The President's Commission on Americans Outdoors; 1986. Reprinted by permission of the author.

Ronald, A. Why don't they write about Nevada? Western American Literature. 24:213–224; 1989. Copyright 1989 by Western American Literature. Reprinted by permission of Western American Literature.

Williams, T. T. Burrowing Owls. In: Williams, T. T. Refuge: An unnatural history of family and place. New York: Pantheon; 1991. Copyright 1991 by Pantheon Books. Reprinted by permission of Pantheon Books.

Index

Abbey, Edward, 73, 74, 76; on Glen Canyon, 98–99, 100; on Wallace Stegner, 64

Aboriginal people: Anasazi, 185–86, 193; Bedouin nomads, 184; and environmental degradation, 184–89; and fire, 189, 192–93; hunting strategies, 185, 188–91; Iticoteri tribe, 156; mythology, 9–10; as noble savage, 184–85, 187, 194–95. *See also* Hunter-gatherers; Indians

Acid rain, 272; industrial responsibility for, 26

Activism. *See* Advocacy, environmental

Adirondack Park, 16, 30; reintroduced species in, 16, 24

Adventures of Huckleberry Finn, 69–70, 73

Advocacy, environmental: and ecological economics, 216; for ecological literacy, 211; for wilderness restoration, 18, 27. *See also* Environmental education

Aesthetics: economics of, 105; soci-

ety's choices for, 147n.16; in urban buildings, 177; of wilderness preservation, 173; and wilderness restoration, 18, 20; and wilderness writing, 107. *See also* Art

Africa, 156–58

Agrarianism, 53

Agriculture: of Anasazi Indians, 185–86; dawn of, 10–11; and habitat destruction, 118; and monoculture, 60; and wilderness, 74. *See also* Civilization

Air quality: bargaining solutions for, 225–27; and economics, 217, 221, 223; of Grand Canyon, 228, 229; and pollution credits, 225–26; subsidies for, 225. *See also* Economics; Pollution

Alaska: Alaska National Interest Lands Conservation Act, 3, 264; Arctic National Wildlife Refuge, 20, 264–65; bears of, 83, 85–87, 91–93; Bob Marshall in, 139–40, 141; grizzly bears of, 82–83, 95

All-terrain vehicles, and environmental degradation, 274, 276

American Childhood, An, 72
American Cottage Tradition, in
pastoral literature, 61
Anasazi. *See* Aboriginal people
Animals: and automobiles, 56; of
Central Park, 176; extinction and
habitat preservation, 4, 16, 21–
22; and genetic engineering, 60; in
natural communities, 18; rights of,
23–24, 31 n.7
Anthropology: of early humans, 186–
87; and evolutionary ecology,
194
Archeology: and creative process,
10; in Dead Sea Valley, 184; game
abundance data, 192; Gatecliff
Shelter site, 190; Olsen-Chubbuck
site, 188; Sudden Shelter site, 190
Arctic National Wildlife Refuge: and
oil development, 20, 264–65. *See
also* Alaska
Arizona, grizzly bears in, 94
Art: and artists, 80; and Jews, 136,
145 n.6; in modern society, 13;
and nature, 4, 10, 54, 76, 173–74;
in pastoral literature, 57; and science, 80, 90–91; urban wilderness
simulations, 174
Atlantic flyway: and birds of Central
Park, 175. *See also* Birds; Wildlife
Automobile: and road kills, 56; and
wilderness access, 238, 246–47,
272

Backpacking. *See* Hiking
Barbarians, modern vision of, 7–8, 14
Basin and Range, 106–7
Bear River Migratory Bird Refuge,
117, 118
Bears: and people, 83, 90–92; in
poetry, 83–85, 87–90; in scientific
study, 91; as symbol of wilderness,
91, 93. *See also* Grizzly bears
Bedouin nomads. *See* Aboriginal
people
Berry, Wendell, 68, 77
Bighorn sheep, 115
Biocentrism, in wilderness manage-
ment, 247

Biodiversity: and habitat preservation, 4, 18, 21, 205, 261. *See
also* Habitat destruction; Species
extinction
Biological knowledge: of grizzly
bears, 81; and species extinction,
21; and wilderness restoration, 19,
29. *See also* Ecology; Science
Biology: and field biology, 81–82;
and "new biology," 47, 48. *See also*
Botany; Scientific inquiry
Birds: Bear River Migratory Bird
Refuge, 117–18; in Central Park,
175; in evolutionary creation, 43;
in Yellowstone Park, 115. *See also*
Wildlife; *and specific topics, e.g.,
Hawks*
Bison: in early Indian hunts, 188–
89; in prairie communities, 118; in
Yellowstone Park, 115–16. *See also*
Animals; Wildlife
Black Elk Speaks, 9
Black-footed ferret, 118, 121
Blixen, Karen. *See* Dinesen, Isak
BLM. *See* Bureau of Land Manage-
ment
Blueprint for the Environment, 220,
234 n.4
Bob Marshall Wilderness: grizzly
bears in, 95. *See also* Marshall,
Bob
Botany: Thoreau's contributions
to, 44, 50n.6. *See also* Biology;
Scientific inquiry
Bulldozer, as literary symbol, 54, 58,
63
Bureau of Land Management (BLM):
timber sale policies, 216; Utah EIS,
149 n.27; and wilderness designa-
tions, 4. *See also* Government; U.S.
Forest Service
Burrowing owls, 117–21

California: grizzly bears in, 94; and
Nevada, 107; trout in, 31 n.5
Camping: and self-discovery, 8,
148 n.19; and travel cost methodology, 229–30; and wilderness

ing in, 61; native plants of, 59; and prairie communities, 118. *See also* Ecosystems; Western states

Desertification, in Near East, 193

Desert Solitaire, 98–99

Devall, Bill, 69, 167, 168

Development, 20, 28: industrial, 161; in pastoral literature, 53–58; in Utah, 118, 184. *See also* Habitat destruction; Land use

Dillard, Annie, 72, 74, 75, 80, 91

Dinesen, Isak, 156–57

Direct experience. *See* Personal experience

Diseases, and Native Americans, 191–92

Donner, Florinda, and *shobono*, 153, 156

Douglas, David, 100, 108 n. 1

Dreaming Big Wilderness, 68

Dreams, 10; of bear attack, 93; and Freud, 13; and law of nature, 64; of pastoral life, 62

Drengson, Alan, 67

Eagles, bald, in Adirondack Park, 16

Early man. *See* Aboriginal people

Eco-feminism. *See* Feminism

Ecological Economics, 214

Ecological literacy: for adults, 210; recommendations for, 211–13; for young people, 207–8, 276–77. *See also* Environmental education

Ecology, 38–39, 47–48, 163; and deep ecology, 168; and *Ecological Economics*, 214; and environmental education, 207–8; in literature, 53–54, 57–60, 63–64, 87; of Thoreau, 42, 45–47, 48 n. 3; and wilderness restoration, 17, 28; for wildlife management, 217–18, 244, 246. *See also* Community; Ecosystems; Niche

Economics, 60, 164, 222–23, 226; and collective ownership of nature, 223, 230–33; and *Ecological Economics*, 214; of government timber sales, 215; and Indian conservation ethic, 189; in Muir's Nevada essays, 105; opportunity cost concept, 218–21, 223, 234 n. 8; of

species reintroduction, 24–25; and wilderness, 105–7, 215, 262; of wilderness recreation, 237, 240, 263–64; of wilderness restoration, 24, 26–27; and willingness to pay, 228–29. *See also* Efficiency; Values

Eco-philosophy, and feminism, 162–66

Ecosophy, and deep ecology, 167

Ecosystems: in Central Park, 175; early human impacts on, 185–91; human impacts on, 21, 184, 188–90, 192–94; restoration of degraded, 16–19, 21–22, 27–30, 31 n. 1; and values, 21, 22, 25, 31 n. 6. *See also* Environmental degradation; Habitat destruction

Eden (biblical), 58, 61, 64, 173

Edom, 183

Education: in North America, 208; practical definition of, 209; in wilderness management, 241–42, 248–49. *See also* Environmental education

Efficiency: and choices for resource use, 219–20; and present value criterion, 215. *See also* Cost-benefit analysis; Economics

Eiseley, Loren, 72–73, 76

Elk: in Yellowstone Park, 115, 116, 192. *See also* Wildlife

Emerson, Peter, 215, 233 n. 1

Emissions. *See* Industrial emissions

Energy policy, and economic policy, 220

Environment: clean-up for, 162; and human nature, 11; and species preservation, 21–22. *See also* Ecosystem

Environmental activism. *See* Advocacy, environmental

Environmental amenities, and environmental economics, 216–17

Environmental degradation: from backpacking, 26, 236, 239, 241–42; by aboriginal cultures, 184–90, 192–94; by all-terrain vehicles, 274; in Central Park, 175; contemporary solutions for, 26–28, 195–96; and ecological concerns, 160–61; economic policies for,

Environmental degradation (*cont'd*) 214; in Near East, 184, 193–94; and relationship with nature, 28–30, 163, 166, 169, 207; and wilderness roads, 216, 219, 246–47, 267. *See also* Fire; Habitat destruction; Pollution

Environmental education, 24, 87, 205–13; and Project WILD, 207–8; for public schools, 207–8, 276–77; and wilderness management, 5, 241–42, 248–49; for wilderness values, 176, 273, 276–77. *See also* Education

Environmental ethics: and animal rights, 23–24, 31n.7; and eco-philosophy, 166; of native Americans, 188–89; and social commitment, 27. *See also* Values

Environmental management. *See* Wilderness management

Environmental quality: cost-benefits of, 215, 217–19, 228–29, 232–33; government standards for, 224–25; market failure and externalities, 222–23; social choices for, 219, 221, 232–33; and willingness to pay, 228–29

Erosion. *See* Soil erosion

Ethics: for environmental education, 209; for environmental protection, 162–64; of wilderness restoration, 16–17, 20. *See also* Rights; Values

European colonization: and game populations, 191–92; and Indian populations, 191; and wilderness destruction, 237. *See also* Aboriginal people

Evergreen trees, in urban landscape, 178

Evolution, 9–12; spiritual, 43, 59

Eyvindur, 128–31

Falcons, in Adirondack Park, 16

Farming: African, 156–58; and genetic engineering, 60; and indigenous species, 24; irrigation for, 61; prairie, 53–54. *See also* Agriculture

Fear of wilderness: American legacy of, 205; in cities, 177, 179

Feminism: and exploitation of environment, 5, 32n.9, 119–21, 162–64, 169. *See also* Rights; Values; Women

Fiedler, Leslie, 143n.2

Fire, aboriginal use of, 189, 192–93

Fishing: economics of, 263–64; and modern society, 272; and wilderness management, 237; in Yellowstone Park, 113–14. *See also* Hunting; Recreation

Flow and movement, in river writing, 73, 74–76

Flow of the River, The, 72–73, 76

Food: and dawn of agriculture, 10–11; in Native American diet, 185–92. *See also* Agriculture; Hunting

Ford Foundation building, 177

Foreman, Dave, 68

Forests: in Central Park, 175; and fire suppression, 193; of Nevada, 105; of South America, 156; tropical, 4, 20–21, 29, 261–63; in urban landscape, 178

Forest Service. *See* U.S. Forest Service

Freedom: river as symbol of, 69–70; in wilderness experience, 68, 153; in wilderness writing, 102

Fremont, John Charles, 102–3

Friendship: sacred objects for, 123; in Thoreau's writing, 74

Fruit tree, as pastoral symbol, 54

Future generations. *See* Obligation

Gardens: and middle-landscape, 173; and natural landscape, 55, 63–64; in pastoral literature, 53, 58–59. *See also* Agriculture; Farming; Landscape

Garden West, in Stegner's writings, 61, 62

Gatecliff Shelter, 190

Genetic engineering, and modern farming, 60

Genetics: and human behavior, 4, 11; and native plants, 59

Geography of hope, in wilderness, 64

Geology, and rivers, 73–74

Ghost towns, in Great Basin, 104

Human beings (*continued*)
184–85, 272; in wilderness management, 240–41. *See also* Women
Humility: in wilderness experience, 139–40; in wilderness restoration, 29–30
Hunter-gatherers: Ache' of Paraguay, 194; myth and religion of, 9–10; as significant predators, 188–91; Yanomamo Indians, 189. *See also* Aboriginal people; Predators
Hunting, 3, 10, 14, 272, 276; in Africa, 158; by Native Americans, 185, 188–91; economics of, 263–64; of grizzly bears, 81–83, 88; and species reintroduction, 24, 32n.8. *See also* Recreation

Iceland, 122–31
Id, in modern society, 13–14
Idaho: mountain lion in, 262; recreational wilderness use, 263–64
Imagination: grizzly bears in, 90; in wilderness writing, 103–4
In a Different Voice, 170n.2
Indians: Anasazi of Utah, 185–86; and bears, 83, 86, 87, 94; conservation ethic of, 187, 189–91; and deep ecology, 168; environmental impacts of, 185–91; and epidemics, 191–92; food resources of, 185–89, 192; of Nevada, 105; value system of, 194. *See also* Aboriginal people
Individual: responsibility for wilderness, 248; and wilderness, 74; and wilderness economics, 230–33. *See also* Society
Industrial emissions: bargaining solutions for, 225–27; environmental damage from, 26; and market failure, 223. *See also* Air quality; Pollution
Insects, in wilderness, 124
Interagency Grizzly Bear Study Team, 95–96
Intergenerational contract: for future generations, 19, 31n.3. *See also* Obligation

Irrigation, 61
Iticoteri tribe, 156

Jefferson, Thomas, 173
Jews: qualities of, 145n.6; in western wilderness, 135–43
Jobs: economic analysis for, 218; and subsidies to loggers, 216; in wilderness restoration, 25
Jordan, 183
Journal, of Thoreau, 37–39
Journey: in Icelandic legend, 125; in river literature, 73–74; spiritual, and grizzly bears, 86

K2, and Julie Tullis, 155
Kittredge, William, 86, 93–94, 105–6, 149n.25
Knowledge, 9, 11–12, 42, 77, 166, 168; for environmental analysis, 218, 234n.3; for wilderness restoration, 17, 23–24, 29. *See also* Personal experience; Values
Kobuk River, 91
Kumin, Maxine, 85, 88–90

Lake Champlain, 24
Lake Powell, 99–100
Lampreys, in Lake Champlain, 24
Land: ownership of, 144n.3; protection for, 60–61; respect for, 53–54, 58, 68–69, 162; and wilderness values, 68–69. *See also* Values
Landscape: human alteration of, 184, 188–90, 192–94; monumental structures in, 172; naturalists as observers of, 71, 75; in Nevada writing, 102–6; in office buildings, 177; in park design, 174; in pastoral literature, 63–64; in wilderness writing, 97–107, 122–23. *See also* Environment; Nature; Wilderness
Landscape architecture, 174–81
Land use: in pastoral literature, 53–56, 58; and wilderness advocacy, 14; and wilderness restoration, 19. *See also* Development; Environmental degradation

Mt. Washington, NH, 50n.6
Muir, John, 68, 237; and Bob Marshall, 138; Nevada essays, 103–5, 107; wilderness writing, 77, 97, 103, 104
Multiple-use areas: Adirondack Park, 16. *See also* Land use; U.S. Forest Service
My Journey to Lhasa, 154–55
Myth, 9, 13, 102; of harmony with nature, 195–96; of pastoral life, 61; of people and bears, 83, 87; and Thoreau's writings, 43, 46

Nash, R., 17, 18, 138, 172–73, 238
National Audubon Society, 215; and Central Park, 175; and environmental economics, 228; and Rainey Wildlife Sanctuary, 227
National Elk Refuge, 115
National Environmental Policy Act (NEPA) of 1969, 214
National Forest. *See* U.S. Forest Service; Bureau of Land Management
National parks: and development, 3; humans and reintroduced species in, 24; North Cascades, 238; and wilderness restoration, 244; Yellowstone, 113–16. *See also* Parks
National Wilderness Preservation System, 239, 240; acreage in, 3; and Wilderness Act of 1964, 3, 258
National Wildlife Federation, 5, 215, 262
Native Americans. *See* Indians
Native plants: and exotic plants, 59, 207; in urban landscape, 178–80; for wilderness restoration, 243, 248. *See also* Plants
Natural history, in wilderness writing, 5
Natural History of Massachusetts, The, 42
Naturalists: and grizzly bears, 81–82; on wilderness landscape, 71
Natural man, in literature, 57–58, 69–70
Natural resources: collective ownership of, 223; and environmental

economics, 217, 223; in wilderness management, 237
Natural Resources Defense Council, 215
Natural science, and self-knowledge, 48
Natural state, and wilderness values, 67–69
Nature, 16–17, 43–44, 172, 208, 222; and hunter-gatherers, 188–89; and new biology, 47, 48; in pastoral literature, 54–58, 63–64; and people, 8, 10–11, 40, 162–63, 272; and urban park design, 174. *See also* Environment; Landscape; Wilderness
Nature Conservancy, The, 227, 275
Nature vs. nurture, and human behavior, 11
Nature writing, 99–102, 107; origins of, 97; science and art in, 94; and self-discovery, 100–101. *See also* Literature; Poetry
Near East, 184, 193–94
Negro Speaks of Rivers, The, 71
Nevada: archeological sites, 190; wilderness writing about, 102–8, 108n.2. *See also* Deserts; Western states
New biology, nature and spirit in, 47–48
New Mexico: Anasazi culture of, 193; Gila National Forest, 3; grizzly bears in, 94
New York City: Central Park, 174–77; naturalist writer in, 125–26; South Cove Park, 179–80; Time Landscape, 178; wilderness simulations in, 174
New York State, Adirondack Park, 16
New York State Department of Environmental Conservation, 16, 24; wolf reintroduction, 32n.8
Niche: definition, 23; of lichens, 44; organism reintroduction into, 19, 23; in Thoreau's *Journal*, 39. *See also* Community; Ecosystems
Nile River, 75
Nomads: myth and sacredness in, 9–10

North America: environmental education in, 208–9; wilderness psyche of, 205–6
North Cascades National Park, impact of hikers on, 238, 241–43, 248
Nuclear wastes, 160–61
Nurturance: and feminine values, 165, 166. See also Feminism

Obligation: to future generations, 19, 31n.3; for species preservation, 22; for wilderness restoration, 20–23. See also Ethics; Restitution
Observations. See Scientific inquiry
Observer and observed: in arctic experience, 124; and life of Alexandra David-Neel, 154; in river writing, 70, 72–74, 76; in wilderness writing, 101–2. See also Personal experience
Oil industry, and Arctic National Wildlife Refuge, 20
Oil spills. See Pollution
Oliver, Mary, 9, 14
Olmsted, Frederick Law, 174
Olsen-Chubbuck site, 188–89
On the Loose, 8–9
Open space: in municipal regulations, 275
Opportunity cost: and environmental economics, 218–23, 234n.8. See also Economics
Other. See Personal experience
Owls. See Burrowing owls
Owning it All, 93–94, 105–6, 149n.25

Pack-trip hunting. See Hunting
Paracelsus, 164
Parks: Adirondack Park, 16; New York Central Park, 174–77; and quantity controls, 224; South Cove, 179–80; wilderness simulations in, 174. See also National Parks
Pastoral literature. See Literature; Poetry
Patriarchy: and deep ecology, 167–

68; in science, 164, 169. See also Feminism; Male Supremacy
Peak experience. See Personal experience
People. See Human beings; Women
People's Forests, The, 141, 144n.3
Personal experience: of environmental destruction, 120–21; and knowledge, 12, 13, 42, 45; of nature, 72, 80, 207; and peak experience, 137; of wilderness, 68, 228–30, 240, 271; of wilderness Others, 68, 70. See also Observer and observed; Relationship
Pesticides, 58
Philosophia Botanica, 44
Photography: for grizzly bear study, 82; and self-discovery, 13
Physical potential, 10
Physics: and deep ecology, 168; natural basis for, 47
Planning. See Wilderness management
Plants, 4, 9, 18, 44; in Iceland, 124; medicinal, 262–63; native vs. exotic, 59; in urban landscaping, 176–77. See also Ecosystems; Native Plants; Vegetation
Platte River, 72
Poetry: about bears, 83–85, 87–90; about the natural state, 83–84; and scientific truth, 42–43, 47; wilderness and river poetics, 73–77. See also Literature; Nature writing
Polarity: and natural truth, 39–40, 42–43, 47–48; of nature and technology, 54–58; in Western values, 164
Pole of Relative Inaccessibility, 127
Politics: of Bob Marshall, 138, 141–42, 144n.5; and environmental standards, 224–25; in pastoral literature, 55; and wilderness management, 239, 245–46; of wilderness preservation, 28, 173, 195–96, 265. See also Government
Pollution: acceptable levels of, 218, 224; acid rain, 26; bargaining solutions for, 225–27; and economics, 222–23; legal suits for, 161; oil

Green, 74; Kobuk, 91; Mississippi, 69–70; Nile, 75; Oldman, 73; Platte, 72; Salmon, 91; Tuolumne, 104; Yellowstone, 114; and Chief Seattle, 71; as symbol of freedom, 69–70; as symbol of rejuvenation, 70, 73; in literature, 73–74
Roads: in wilderness areas, 216, 219, 267; in wilderness management, 246–47. *See also* U.S. Forest Service
Rock climbing, 8
Rolston, H., 20, 22, 25, 31 n.4
Romanticism, 173–74
Running, 12
Run, River, Run, 75

Sacredness: of agriculture, 11; of grizzly bears, 114–15; of objects, 123; and sacred tree, 9; of wilderness space, 68; of Yellowstone Park, 114
Sailing, 8
Salmon, in Lake Champlain, 24
Salmon River, 91
Salt blocks, for wildlife management, 243
Sanity: in natural wilderness, 68; and natural wisdom, 42, 46; and subordination of feminine, 162–63. *See also* Health
Saskatchewan, 53
Schools. *See* Education; Environmental education
Schultheis, Rob, 100, 108 n.1
Science, 20, 42–43, 90, 208; and global catastrophe, 160; and patriarchy, 164–65, 169; and scientists, 80–81. *See also* Technology
Science, Technology, and Society (STS), 208
Scientific inquiry, 13, 42, 81; grizzly bear study, 81–82; and literature, 90, 94; and "new biology," 47; and personal experience, 45; for wilderness preservation, 268–69. *See also* Research
Scientific positivism, 42–43, 47–48
Seasons: and burrowing owls, 117; in

river literature, 73; in Yellowstone Park, 113–16
Seattle, Chief, on rivers, 71
Self-discovery, 8–9, 83–85; in grizzly bear hunt, 86–87; and "wilderness man," 67; in wilderness writing, 38, 41, 74, 100–101. *See also* Personal experience
Service, Robert W., 258–59
Seven Rivers of Canada, 74–75
Shooting Star, A, 54–55, 57–58, 64
Sierra Club, 215
Sierra Nevada, 97; in Muir's writings, 104
Skiing, 8
Slash-and-burn. *See* Deforestation
Snyder, Gary, 87–88
Society: contemporary, 10, 21–22, 260; and economics, 217–21; evolution of, 8–13; and government, 12, 209–10, 269, 277; and individual, 230–33; and pollution, 161, 223; and subordination of feminine, 162–63; and wilderness, 7–9, 27, 68–69, 173, 206–7, 229–30. *See also* Human beings; Individual
Soft animal, and human spirit, 9, 14
Soil erosion: and Anasazi culture, 193; from wilderness hikers, 238
Sound of Mountain Water, The, 56, 99
South Cove Waterfront Park, 179–80
Space: in arctic, 123; sacred, 68
Space program, 25–26
Species: ecology of, 261–62; indigenous, 23–24; inherent value of, 21–22, 31 n.6. *See also* Animals; Niche; Plants
Species extinction: by aboriginal Americans, 189–90; by aboriginal Polynesians, 193–94; and species reintroduction, 24; and wilderness corridors, 21, 29; and wilderness management, 239, 275; world species populations, 261–63. *See also* Ecosystem; Habitat destruction
Spectator Bird, The, 58, 59–60
Spirit. *See* Psyche

Stegner, Wallace, 52–65, 99, 135, 143n.2
"Stickers," and wilderness immigrants, 143n.2, 146n.13
Story From Bear Country, 83–84
Subsidies: for environmental quality, 225; to timber industry, 215–16. *See also* Government
Suburbs, in Stegner's writings, 58
Succession (ecological), in urban landscape, 178
Sudden Shelter, 190
Survival. *See* Species extinction
Survival, human: instincts for, 9–13; and wilderness values, 206
Symbolic Interactionism, 170n.1
Symbols in wilderness: bears, 91, 93; rivers, 69–70. *See also* Literature

tat twam asi, and interdependence of life, 39
Taxes, for pollution, 225
Technology, 10, 61, 160; in literature, 53–58; in park creation, 175; in wilderness activity, 238, 272, 274; and wilderness restoration, 19–20, 24. *See also* Science
Thoreau, H. D., 4, 11, 12, 37–50, 97; Concord journeys of, 44–46, 49n.5; ecological walks, 38, 41, 44, 45–46, 49n.5; and "ecology," 48n.3; lichen botany, 44–47, 50n.6; observations of nature, 38, 39, 42, 44–46; on rivers, 70, 72, 73, 74, 76; transcendentalism of, 39–40, 42–43; and W. E. Channing, 44, 49n.5
Timber industry: Bob Marshall on, 141, 144n.3, 149n.25; government subsidies for, 215–16. *See also* Logging; U.S. Forest Service
Time Landscape, 178
Tongass National Forest, 4
Tourism: economics of, 263–64; at Yellowstone National Park, 113
Tragedy of the Commons, and natural resource waste, 223. *See also* Economics
Transcendentalism: and river experi-

ences, 71; in Thoreau's writings, 43
Transformation. *See* Self-discovery
Tropical forests: destruction of, 20, 261; medicinal plants in, 262–63; restoration of, 29; species extinction in, 21. *See also* Forests; Habitat destruction
Trout: reintroduction of, 31n.5; as symbol of rejuvenation, 70; in Yellowstone Park, 113
Truth: and dualism in Western thought, 164; and transcendentalism, 42–43
Tullis, Julie, 155–56
Tuolumne River, 104
Twain, Mark, 69–70, 73

U.S. Fish and Wildlife Service, 24
U.S. Forest Service: and Bob Marshall, 136, 141; Limits of Acceptable Change planning, 244; timber sale policies, 215–16; wilderness management policies, 239–42, 245, 246. *See also* Bureau of Land Management
Unity, poetic perception of, 77
Urban civilization: and western Jews, 136–38, 143–45n.2–6. *See also* Cities
Utah: archeological sites, 185–86, 190; grizzly bears in, 94; habitat destruction in, 118, 149n.25, 184. *See also* Aboriginal people; Western states

Values: in conquest of nature, 162–65; educational, 208–9, 212; land use, 53–54, 60; shaped by experiences, 194; in urban wilderness simulations, 174; for wilderness, 25, 67–69, 71–72, 164, 206, 228–30; for wilderness management, 245; in wilderness restoration, 16–17, 20–21, 24–25, 27, 31n.1, 31n.4. *See also* Economics
Vedas, and Thoreau, 47
Vegetation, recreational impacts on, 236, 238, 240, 242–43, 248

Vision: in creative imagination, 81; for wilderness management, 245–46. *See also* Imagination

Walden, 37–38, 40, 46, 47. *See also* Thoreau, H. D.

Walking: Bob Marshall's ambition for, 137; in Iceland wilderness, 124

Washington: Greater North Cascades Ecosystem, 236; Mt. Baker-Snoqualmie National Forest, 238

Water: instinct for discovery, 9; in ornamental landscaping, 207

Weather: and Native Americans, 185–86; and wilderness experiences, 139, 148n.20; in Yellowstone Park, 113–16. *See also* Climate

Week on the Concord and Merrimack Rivers, A, 37–38, 70, 72–74, 76. *See also* Thoreau, H. D.

Western states: game abundance, 192; native plants, 59; in pastoral vision, 53, 61–63

Wilderness, 17, 20, 74–75, 77, 172, 195–96, 206; and agriculture, 74; and art, 4, 10, 54, 76, 173–74; benefits of, 20, 173, 205; in cities, 5–7, 172–81; and civilization, 8, 54–56, 67–68, 101, 173, 272–73; contemporary concepts of, 157, 187, 205–6; death in, 90, 116; and economics, 105, 216, 224; and environmental education, 205–13, 276–77; and feminism, 5, 32n.9, 119–21, 162–64, 169; and fire, 192–93; human impacts on, 17, 183–92, 195, 236–38, 240–43, 272–73, 276; and "pristine" harmony, 185, 187–88, 194–95, 276; and spirit, 12, 14, 64, 67, 205. *See also* Personal experience

Wilderness Act of 1964, 214, 258; and National Wilderness Preservation System, 3; wilderness definition in, 237

Wilderness corridors: for species survival, 21, 29. *See also* Species extinction

"Wilderness man." *See* Self-discovery

Wilderness management: and biocentrism, 247; and ecological economics, 215–18, 227–31; in Greater North Cascades Ecosystem, 236, 239, 246; information and education (I&E) divisions, 277; Limits of Acceptable Change planning, 244; and management problems, 240–42, 245–46; and people management, 240–41; planning for, 244–45; and wilderness restoration, 243–44, 262; and wilderness roads, 246–47; zoning for, 276. *See also* Recreation

Wilderness restoration, 18, 28, 173, 187; in Adirondack Park, 16; costs and benefits, 20, 25–26; ethics and values of, 16–17, 20–21, 24–25, 27, 31n.1, 31n.4; and hubris, 28–30; hypothetical ecosystems for, 187; in marshlands, 20, 175; moral imperative for, 17, 30, 31n.1; native plants for, 243, 248; and wilderness preservation, 4, 16–19, 22, 27, 30, 268. *See also* Conservation; Obligation

Wilderness Society: and Bob Marshall, 137, 138, 142; and ecological economics, 215; and wilderness management, 244

Wilderness Sojourn: Notes in the Desert Silence, 108n.1

Wilderness values, 67–69, 176. *See also* Values

Wilderness writing. *See* Nature writing; Literature

Wildlife: in Central Park, 176; contemporary ethic for, 272–74; and environmental economics, 217; environmental education about, 207; and monoculture, 60; and quantity controls, 224; reintroduction of, 16, 19, 23–24, 29, 31n.1, 31n.5, 31n.7, 275–76; and wilderness corridors, 21, 29; and wilderness management, 240, 243, 266, 275; in Yellowstone Park, 114–16. *See also* Animals; *and specific topics, e.g. Wolf*

Wildness: and human beings, 12–13,